Computational Biomechanics for Medicine

Karol Miller · Adam Wittek · Martyn Nash ·
Poul M. F. Nielsen
Editors

Computational Biomechanics for Medicine

Solid and Fluid Mechanics Informing Therapy

 Springer

Editors
Karol Miller
Intelligent Systems for Medicine
Laboratory
The University of Western Australia
Perth, WA, Australia

Adam Wittek
Intelligent Systems for Medicine
Laboratory
The University of Western Australia
Perth, WA, Australia

Martyn Nash
Auckland Bioengineering Institute
University of Auckland
Auckland, New Zealand

Poul M. F. Nielsen
Auckland Bioengineering Institute
University of Auckland
Auckland, New Zealand

ISBN 978-3-030-70125-3   ISBN 978-3-030-70123-9   (eBook)
https://doi.org/10.1007/978-3-030-70123-9

This Springer imprint is published by the registered company Springer Nature Switzerland AG
The registered company address is: Gewerbestrasse 11, 6330 Cham, Switzerland

Preface

Through prediction of biomechanical responses of human body organs, tissue rupture risk, and quantitative understanding of physical phenomena that govern disease mechanisms and disease progression, computational biomechanics simulations provide unprecedented opportunities to extend medical professionals' ability to diagnose, plan and carry out treatment more effectively and with less trauma to a patient. While in research laboratories, substantial advancements towards clinically relevant computational biomechanics algorithms have been achieved using models and simulations, there is still much work ahead before personalised medicine, underpinned by personalised computer simulations, becomes an integral part of healthcare.

The first volume in the Computational Biomechanics for Medicine book series was published in 2010. Since then, the book has become an annual forum for specialists in computational sciences to present their latest results and discuss the opportunities and challenges of applying their techniques to computer-integrated medicine. This twelfth volume in the Computational Biomechanics for Medicine book series comprises the latest developments in solid biomechanics, vascular biomechanics, tooth biomechanics, childbirth biomechanics, spinal cord biomechanics and machine learning from researchers from Australia, Czech Republic, Denmark, France, Germany, New Zealand, Serbia and USA. Some of the topics discussed include:

- Frameworks for computer-assisted therapy;
- Disease and injury mechanisms;
- Biomechanical tissue characterisation;
- Determining organ geometry from images;
- Organ deformation measurements.

Computational Biomechanics for Medicine books continue to provide the most up-to-date source of information for researchers and practitioners alike.

Perth, Australia   Karol Miller
Perth, Australia   Adam Wittek
Auckland, New Zealand   Martyn Nash
Auckland, New Zealand   Poul M. F. Nielsen

Contents

Keynote Abstract

From Simulation-Based Design to Simulation-Based Treatment
The work summarizes the opportunities, which the virtual world could contribute to the real world. The virtual biomechanical human body models are discussed as the major tool in the virtual approach in biomedical engineering. Simple articulated rigid body-based human models as well as detailed deformable finite element models are discussed. Biofidelity, validation, and injury assessment of both approaches are explained. Due to their simplicity, the articulated rigid body-based models calculate very fast compared to the detailed finite element models. On the other hand, whilst the articulated rigid body models bring just basic information on injuries based on the criteria, the finite element-based models can describe the particular injury including the injury mechanism. Therefore, the hybrid models are also discussed as the compromise between both approaches. They benefit from the fast calculation thanks to the basic articulated body structure and the injury description defined using the deformable segments. The exploitation of a hybrid model for simulation-based design in engineering with a special focus on vehicle safety is presented. Scaling method for developing a wide spectrum of personalized anthropometric human body model based on a reference model is shown. The presented work further shows the procedure, how the engineering approach using the person-specific human body models can help the medical world by diagnoses, prosthesis design, or operation planning. The step from the simulation-based design to the simulation-based treatment is shown using a detailed pelvic floor model to be used in birth injury prevention. The particular example concerns the prediction of the successful course of vaginal delivery in the relation of bony pelvis anatomy and fetal head size and position based on a female-specific pelvic floor model. Using the presented model composed of bones, muscles, and skin, the strains during the second phase of delivery leading to the perineal loading are assessed. The work establishes such approaches for future use in industry and human care.

Luděk Hynčík
University of West Bohemia
Pilsen, Czech Republic
hyncik@ntc.zcu.cz

Computational Biomechanics Frameworks and Models for Computer-Assisted Therapy and Understanding of Disease Mechanisms

Automatic Framework for Patient-Specific Biomechanical Computations of Organ Deformation

Saima Safdar, Grand Joldes, Benjamin Zwick, George Bourantas, Ron Kikinis, Adam Wittek, and Karol Miller

Abstract Our motivation is to enable non-specialists to use sophisticated biomechanical models in the clinic. To further this goal, in this study, we constructed a framework within 3D Slicer for automatically generating and solving patient-specific biomechanical models of the brain. This framework allows determining automatically patient-specific geometry from MRI data, generating patient-specific computational grid, defining boundary conditions and external loads, assigning material properties to intracranial constituents and solving the resulting set of differential equations. We used Meshless Total Lagrangian Explicit Dynamics Method (MTLED) to solve these equations. We demonstrated the effectiveness and appropriateness of our framework on a case study of craniotomy-induced brain shift.

Keywords Patient-specific modelling (PSM) · Non-linear computational biomechanics · Brain · Brain shift · Automated computations

1 Introduction

We are at the verge of a new exciting era of personalized medicine based on patient-specific scientific computations. These computations usually involve solving models described by boundary value problems of partial differential equations (PDEs). The most common and useful are models of biomechanics, bioheat transfer and bioelectricity.

S. Safdar (✉) · G. Joldes · B. Zwick · G. Bourantas · A. Wittek · K. Miller
Intelligent Systems for Medicine Laboratory, The University of Western Australia, Perth, Western Australia, Australia
e-mail: saima.safdar@research.uwa.edu.au

R. Kikinis
Brigham and Women's Hospital, Harvard Medical School, Boston, MA, USA

K. Miller
Harvard Medical School, Boston, MA, USA
e-mail: karol.miller@uwa.edu.au

In this paper, we are especially interested in patient-specific biomechanics as a tool to compute soft tissue deformations for operation planning and intraoperative guidance. While the methods for patient-specific biomechanical model generation [28] and solution [11, 12] exist, they are very sophisticated and require very high level of specialist expertise from the users. Therefore, the objective of the work described here is to create an automatic framework so that these sophisticated computations can be conducted in the clinic by a non-specialist.

We integrated the framework to automate the process of generating and solving patient-specific biomechanical models into 3D SLICER (http://www.slicer.org/), an open-source software for visualization, registration, segmentation and quantification of medical data developed by Artificial Intelligence Laboratory of Massachusetts Institute of Technology and Surgical Planning Laboratory at Brigham and Women's Hospital and Harvard Medical School [6].

We demonstrated the application of our framework using a case study of craniotomy induced brain shift obtained from our collaborator's database (Computational Radiology Lab, Harvard Medical School), used by us previously [7, 23, 30]. The paper is organized as follows: In Sect. 2, we presented the proposed framework. In Sect. 3, we showed our results based on the case study. Section 4 contains discussion and conclusion.

2   Proposed Framework

The four main steps of the proposed framework workflow (see Fig. 1) are as follows:

1. Image Pre-processing
   Determining patient-specific geometry from medical images
2. Model Construction
   Patient-specific computational grid generation
   Defining boundary conditions and external load
   Assigning patient-specific material properties to brain tissues
3. Model Solution
   Computation of tissue deformations using Meshless Total Lagrangian Explicit Dynamics Algorithm (MTLED)
4. Image Warping, using the computed deformation field

The details of each step are given in Sects. 2.1–2.3. Image warping is done with example case study under Sect. 3.

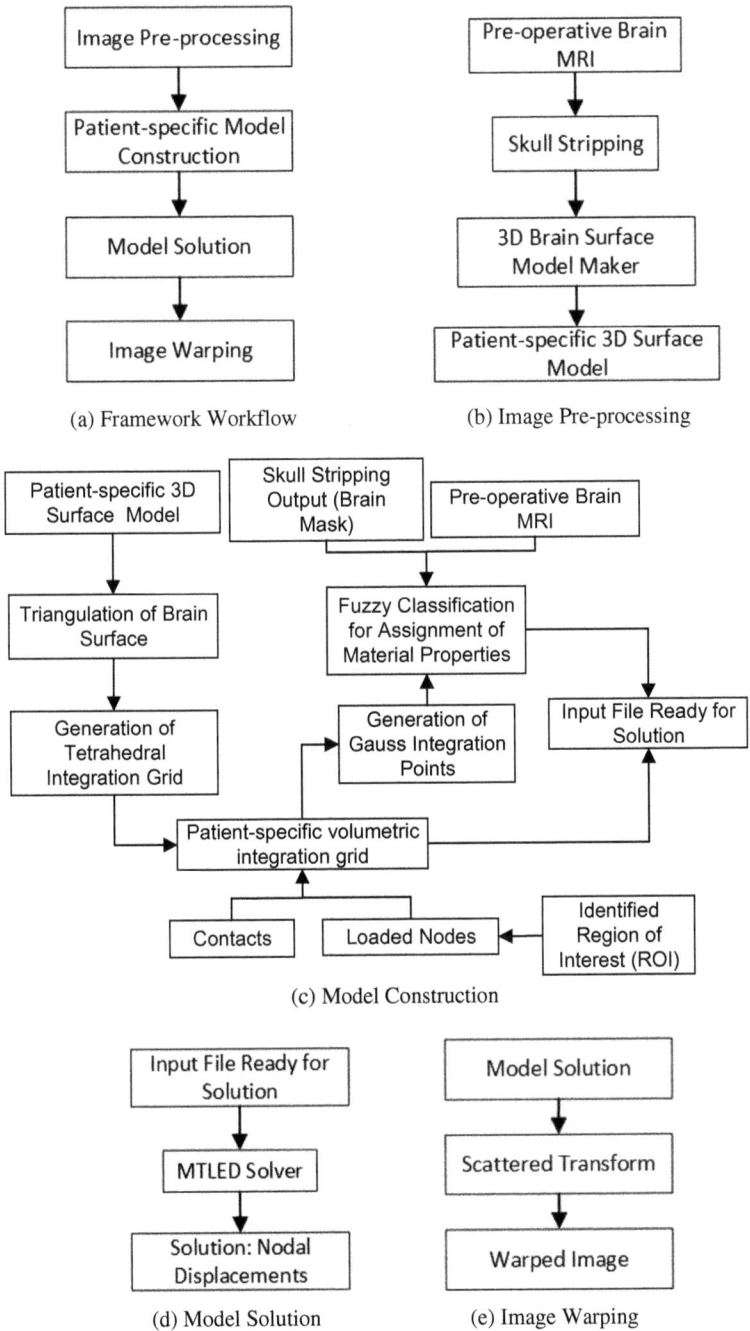

(a) Framework Workflow

(b) Image Pre-processing

(c) Model Construction

(d) Model Solution

(e) Image Warping

Fig. 1 Workflow diagram for patient-specific biomechanical interpretations of organ deformations

2.1 Image Pre-processing

2.1.1 Determining Patient-Specific Geometry from Medical Images

To obtain the geometry of the brain, the skull needs to be removed from the preoperative MRI image. We remove the skull and extract the brain volume using FreeSurfer software (http://surfer.nmr.mgh.harvard.edu/) (see Fig. 2). It is an open source software suite for processing and analyzing human brain medical resonance images (MRIs) [4]. We wrote Python-based scripted modules within 3D Slicer to execute all the remaining steps.

After extracting brain volume, we use threshold filter [25] of 3D Slicer to select the brain parenchyma, see Fig. 3a. We created a three dimensional surface model based on the selected region using model maker module of 3D Slicer, see Fig. 3b [17].

Information about the location and extent of craniotomy is necessary for biomechanical modeling of neurosurgery. We segmented the portion of craniotomy region and created a surface model using this segmented portion (see Fig. 4) and used it to select the brain craniotomy region on the brain surface model. We define external loads on the surface of brain selected through craniotomy segmentation.

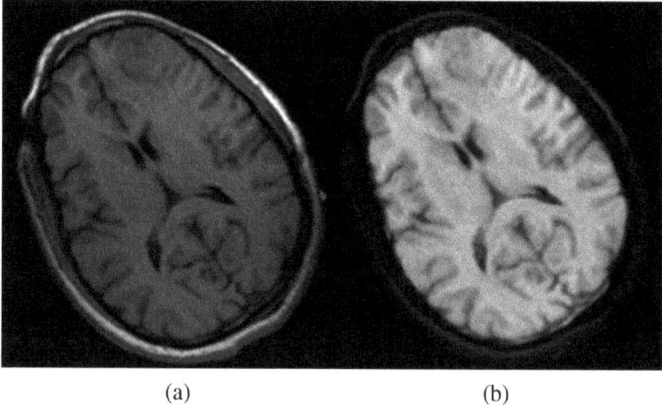

(a)   (b)

Fig. 2 Results of skull stripping for patient-specific preoperative MRI image. **a** Preoperative MRI image, **b** Skull-stripped MRI image aligned with the preoperative MRI image

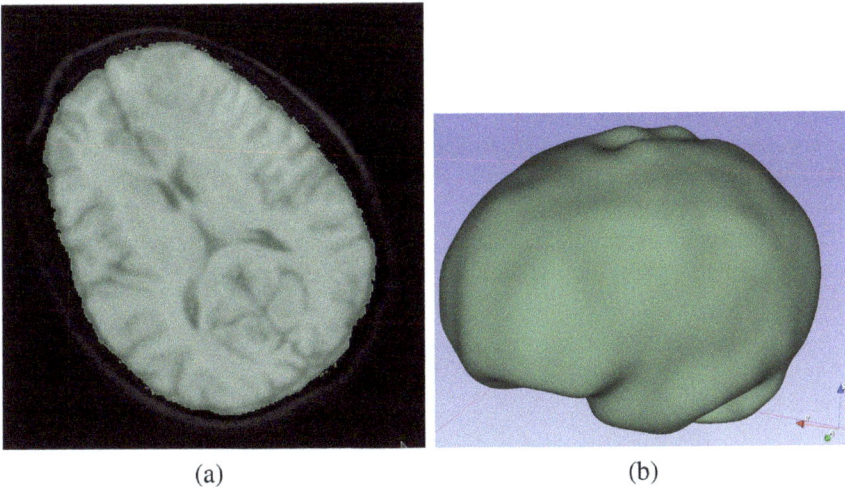

(a)   (b)

Fig. 3 Results of patient-specific brain geometry extraction from FreeSurfer output. **a** Results of threshold filter in 3D Slicer; **b** Surface visualization of the problem geometry produced by surface model maker of 3D Slicer with 45% value of Laplacian filter

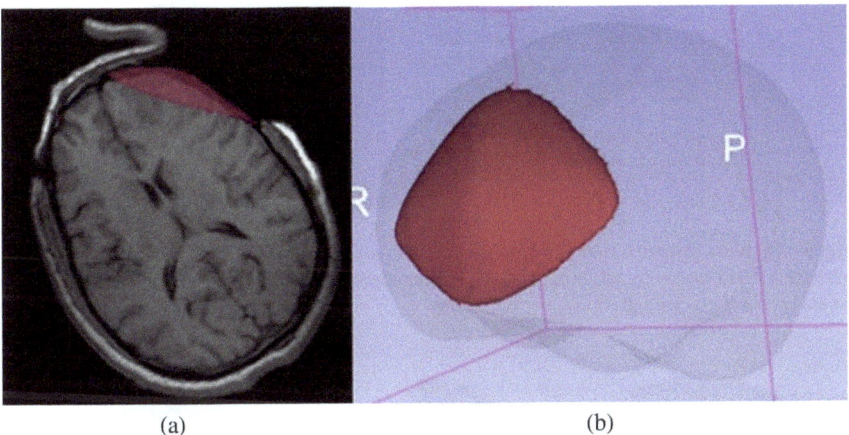

(a)   (b)

Fig. 4 **a** Craniotomy region segment, **b** 3D model of the patient-specific craniotomy region shown together with the entire patient-specific brain model

2.2 Model Construction

2.2.1 Patient-Specific Computational Grid Generation

In our method, we use a tetrahedral background integration grid that conforms to the problem geometry [11]. The volumetric integration (a step in the MTLED solution method, see Sect. 3) is performed over this background integration grid (Gauss integration with four Gauss points per tetrahedron) and displacements are calculated over the cloud of points formed by the nodes of the tetrahedra, Fig. 5. Creating such background grids is fully automatic (i.e. does not require any manual correction). It is very important to note that our tetrahedral integration grid is NOT a finite element mesh and does not need to conform to strict quality requirements demanded by the finite element method.

Our framework uses ACVD (Surface Mesh Coarsening and Resampling) [27] to construct a patient-specific triangulated brain surface (see Fig. 6b) which is then

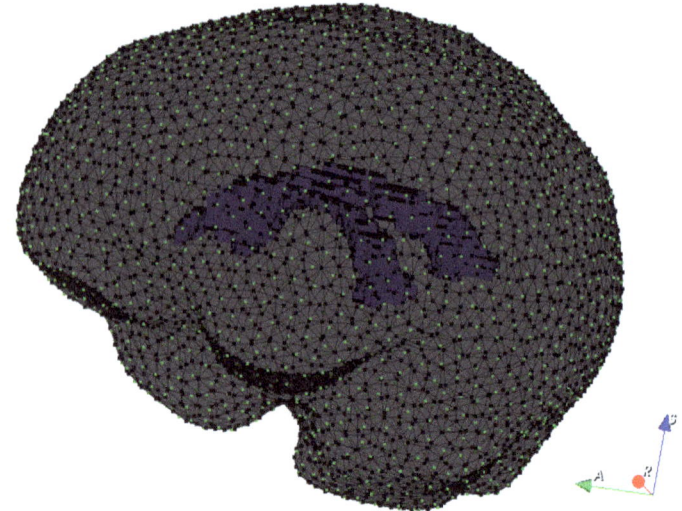

Fig. 5 Meshless discretization for simulation of craniotomy induced brain shift. In this example we have 12,014 nodes (black dots) and 28,915 tetrahedral integration cells with four integration points (green dots) per cell

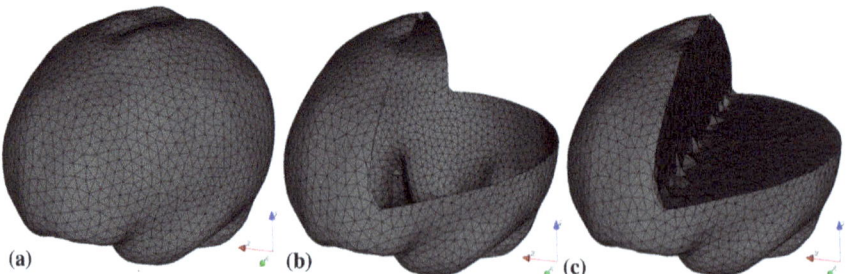

Fig. 6 a Patient-specific tetrahedral integration grid with triangular surface mesh, **b** Example of triangulated patient specific brain surface mesh model, **c** Example of patient specific brain volumetric integration grid filled with tetrahedral cells (geometry conforming tetrahedral cells based biomechanical model)

Fig. 7 Loaded nodes (orange dots) on the surface of the patient-specific brain exposed by the craniotomy

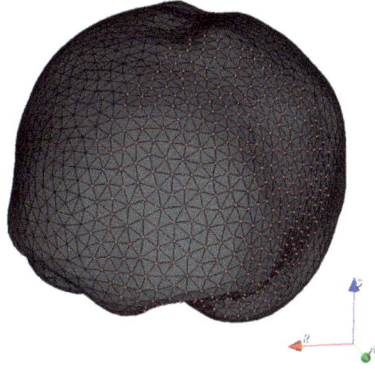

used for generating a 3D integration grid filled with tetrahedral integration cells (see Fig. 6c) using Gmsh [8]. The triangulated surface is also used for defining contacts. All these steps were implemented within 3D Slicer and are automatic.

Similarly, we created a patient specific craniotomy region surface model. We used this craniotomy surface to define the fiducials (3D points in 3D Slicer). We used these fiducials to select the nearest triangulated brain cells exposed to craniotomy region. We applied our nearest neighbor search algorithm in 3D Slicer, which takes into consideration the problem geometry and the craniotomy region fiducials.

We defined loaded nodes (see Fig. 7) based on the selection of triangulated brain cells.

2.2.2 Defining Boundary Conditions and External Load

Boundary Conditions

The stiffness of the skull is several orders of magnitude higher than that of the brain. Therefore, to define the boundary conditions for nodes other than displaced nodes on the exposed surface of the brain, a contact interface is defined between the rigid interface model of the skull and the deformable brain model. Nodes on the brain surface could not penetrate the skull, but could slide without friction or separate from the skull as described in [14].

We created a skull interface using the triangulated surface cells generated as described in Sect. 2.2.1 to define contacts automatically on the surface of patient specific brain biomechanical model.

External Load

Load can be defined either through forces (prescribing natural BCs) or displacements on the boundary (prescribing essential BCs). It is rather difficult to make patient-specific measurements of forces acting on the brain during surgery but there are well-established methods for determining the displacements on the boundaries from images. Furthermore, if we use forces, to accurately compute intraoperative deformations, we need accurate information about patient-specific material properties of the brain tissues. As there is no commonly established method to accurately determine patient-specific material properties of soft tissues from radiographic (MR, CT) images, we define the load through imposed motion of essential BCs. This makes the computed deformations only very weakly dependent on uncertainty in patient specific information about tissue material properties [21, 22, 29].

To define intraoperative loading, intraoperative information is required such as measurement of the current position of the exposed surface of the brain. This can be done through cameras [18] and a pointing tool of a neurosurgical station [26].

2.2.3   Assignment of Patient-Specific Material Properties: Fuzzy Tissue Classification

Material properties of the intracranial constituents are assigned to integration points within the problem geometry through fuzzy tissue classification [1] algorithm. Hard segmentation of brain tissues is difficult to automate [5] and therefore it is incompatible with clinical workflows. Therefore, we integrated a fuzzy tissue classification algorithm [16, 32, 33] into our framework to automatically assign material properties to brain tissues. Slight inaccuracies of tissue properties assignment do not affect the precision of intraoperative displacement prediction because the external load is defined though prescribed essential boundary condition motion rendering the problem Dirichlet-type [21, 32, 33].

In this framework, a neo-Hookean constitutive model (see Table 1) was used for brain tissues and for tumor with Poisson's ratio of 0.49, whereas 0.1 was used for the ventricles [29, 30]. This simple model is used as the simulation belongs to the special class called displacement-zero traction problems (or Dirichlet-type problems) whose solutions are known to be weakly dependent on the unknown patient-specific material properties of the tissues [3, 24, 29].

Table 1 Material properties of biomechanical model

Model components	Density (kg/m^3)	Young's modulus (Pa)	Poisson's ratio
Parenchyma	1000	3000 [19]	0.49 [31]
Tumor	1000	9000 [31]	0.49 [31]
Ventricle	1000	10 [19]	0.1 [30, 31]
Skull	Rigid		

2.3 Model Solution

2.3.1 Computation of Tissue Deformations: Meshless Total Lagrangian Explicit Dynamics Algorithm

MTLED is a numerically robust and accurate meshless algorithm [9, 11]. The method computes deformations at an unstructured cloud of nodes used to discretize the geometry instead of elements as in finite element methods, which requires a high quality mesh of problem geometry [28]. The proposed algorithm uses explicit time integration based on the central difference method. Unlike implicit time integration, this does not require solving systems of equations at every time-step making the method robust in performing calculations [9].

MTLED was evaluated extensively in computing brain deformations on problem geometry based on patient specific MRI data. The simulation results presented were within limits of neurosurgical and imaging equipment accuracy (~1 mm) [11, 20]. The method is also capable of handling very large deformations as well as cutting [10].

Meshless methods are preferred as finite element methods, due to excessive element distortion, are unreliable in scenarios where human soft tissues undergo very large strains in the vicinity of contact with a surgical tool while MTLED gives reliable results for compressive strains exceeding 70% [11].

We developed a separate module which integrates the MTLED solver within 3D Slicer. The MTLED solver uses three input files automatically generated using our framework, which are: (1) computational grid information file, (2) material properties and (3) external load information file. All remaining parameters of MTLED are set by default (see Table 2) and are based on the experience obtained through numerous applications in computing soft continua and soft tissue deformations. The end user can change these parameters as per requirements but we recommend that a non-specialist user leave them unaltered.

Table 2 Default parameters list for MTLED simulator

MTLED parameters	Values
Mass scaling [9]	True
Integration points per tetrahedron [9]	4
Shape Function Type [2]	mmls
Basic Function Type [2]	Quadratic
Use exact derivatives [2]	True
Dilation Coefficient [2]	1.8
Load file curve [9]	Smooth
Node set	Contacts
Surface	Skull
Load time for running simulation	1.0
Equilibrium time [13]	5

3   Craniotomy-Induced Brain Shift Case Study

The image preprocessing was conducted with FreeSurfer and 3D Slicer threshold filtering as explained in Sect. 2.1. The automatic generation of patient-specific computational grid, as explained in Sect. 2.2, resulted in 12,014 nodes and 115,660 integration points (see Figs. 5 and 6c). Contacts are defined as described in Sect. 2.2.2.1. Material properties of brain tissues are defined using fuzzy tissue classification as explained in Sect. 2.2.3.

For the purpose of this case study, we defined external load using a transform obtained through using region of interest (ROI), which is brain craniotomy region in both the preoperative and intraoperative MRI (see Fig. 8a, b). We used crop volume module in 3D Slicer to obtain the preoperative segment (PS) and the intraoperative segment (IS). We used Affine with twelve degree of freedom (12 DOF) and BSpline with greater than twenty seven degree of freedom (>27 DOF) algorithms from 3D Slicer General Registration Module to obtain a transform that aligns PS to IS. We applied this transform on loaded nodes selected on the brain surface to get the transformed nodes (see Fig. 8c). We defined external load based on the information of displaced nodes (loaded nodes) in 3D plane before and after the transform, thus prescribing external load through essential boundary conditions.

We used nodal displacements computed by MTLED (see Fig. 9) in the scattered transform module [15] in 3D Slicer to obtain the transform to warp the preoperative MRI image into the intraoperative configuration of the patient's brain, see Fig. 9a.

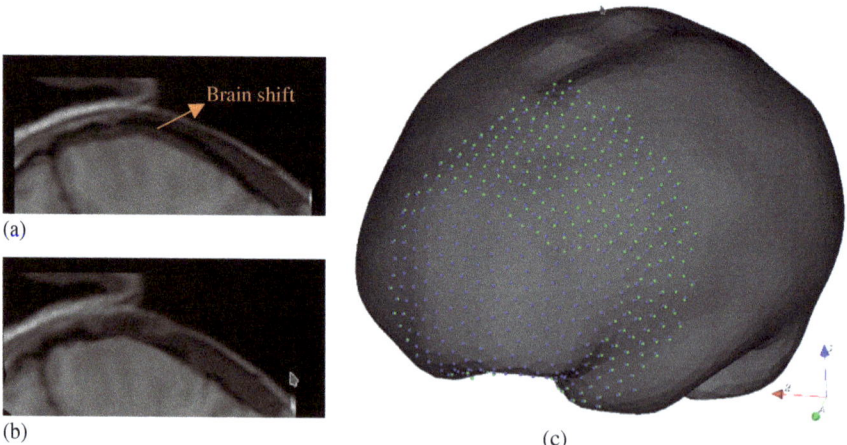

Fig. 8 Affine (12 DOF) and BSpline (>27 DOF) transform results generated by using General registration module within 3D Slicer. **a** Region of interest (ROI) preoperative segment (PS) aligned on top of intraoperative segment (IS) before Bspline transform to show the brain shift (see dark grey area indicated by an arrow, which is the difference between preoperative and intraoperative MRI), **b** results of Bspline transform on images, and **c** Blue dots are pre-transformed fiducials and green dots indicate transformed fiducials obtained from applying the Bspline

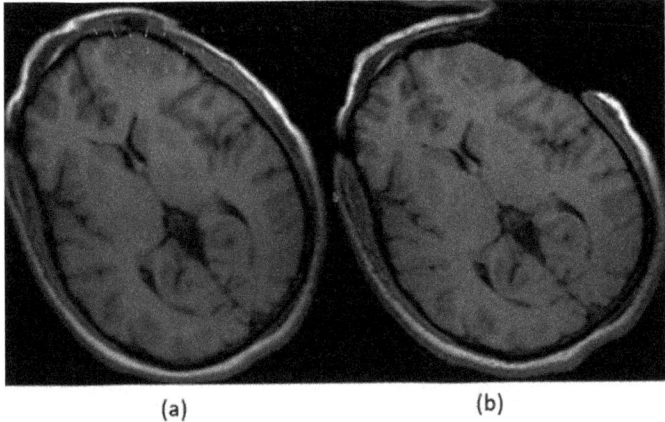

(a)   (b)

Fig. 9 Results of the deformation field extracted from deformed model generated by MTLED. **a** Warped preoperative (Predicted Intraoperative) MRI with deformation field extracted using scattered transform from deformed model generated by MTLED, **b** Intraoperative MRI

We used the warped MRI to identify contours of ventricles and compared them with contours of ventricles from the intraoperative MRI image. We performed ventricle segmentation using a threshold filter [25] and an island selector filter [6] in 3D Slicer for both the warped MRI and the intraoperative MRI. We present ventricle contours comparison without any manual corrections or editing (see Fig. 10). Figure 10 confirms the acceptable accuracy of the presented modelling approach.

The simulation presented in this study was performed on a HP ProBook with Intel Core i7 2.7 GHz processor and 8 GB of physical memory. The calculation time for generating automatically a patient-specific computational model with all details, including patient-specific geometry construction, craniotomy region selection, external loading and defining contacts was 156.87 s. The execution time of the MTLED solution algorithm (i.e. obtaining the deformed model) was 258 s. The time for warping the preoperative image with the deformation field extracted from the model was 0.7 s.

4 Discussion and Conclusion

In this paper we described the framework for automated solution of computational biomechanics problems described by partial differential equations of solid mechanics. We also demonstrated the effectiveness of this framework using a computational biomechanics of the brain example. The framework is integrated within 3D Slicer. It allows automatic generation of patient-specific geometry along with defining the craniotomy region, external load, material properties and boundary conditions.

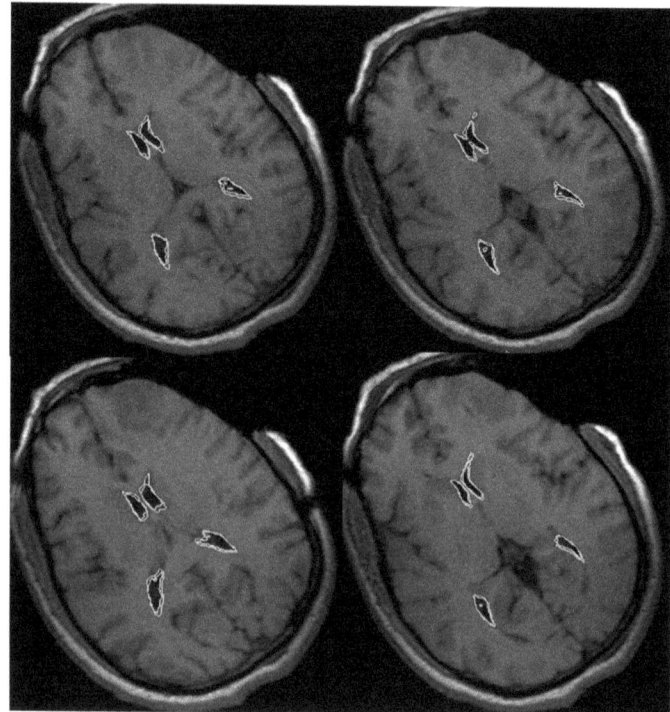

Fig. 10 Intraoperative MRI overlaid with contours (green lines) of the deformed ventricles as generated by MTLED algorithm. The yellow lines represent the ventricle contours of the intraoperative MRI extracted using the threshold filter and the island selector filter in the 3D Slicer

We obtained a solution to this model using an MTLED-based suite of algorithms which is also integrated into 3D Slicer. The results (see Fig. 10) of the automatic simulation show good agreement with the ground truth provided by the intraoperative MRI.

A craniotomy induced brain shift is simulated with 12014 nodes (i.e. ~36000 differential equations are solved) and 28915 integration cells. The patient specific biomechanical model construction, which involves defining the patient specific brain geometry from a preoperative MRI image, patient-specific tetrahedral integration grid generation, defining boundary conditions and external loads, and assigning material properties to brain tissues, took 156.87 s of computer processing time. The solution of the model using our MTLED algorithm took 258 s and finally the image warping took 0.7 s.

These results indicate that the proposed methodology is compatible with clinical workflows and in our future work we will attempt to incorporate in operation planning and neuronavigation systems.

Acknowledgements The funding from NHMRC grants APP1162030; APP1144519 is gratefully acknowledged. The first author acknowledges scholarship funding from University Postgraduate Award. We also wish to thank 3D Slicer on-line community https://discourse.slicer.org/ whose members have made many valuable contributions. Our special thanks go to Dr Andras Lasso of Laboratory for Percutaneous Surgery (PerkLab) at Queen's University (Kingston, Ontario, Canada).

References

1. Bezdek, J. C., Ehrlich, R., Full, W. (1984). FCM: The fuzzy c-means clustering algorithm. *Computers & Geosciences 10*(2,Äì3), 191–203.
2. Chowdhury, H., Joldes, G., Wittek, A., Doyle, B., Pasternak, E., & Miller, K. (2015). Implementation of a modified moving least squares approximation for predicting soft tissue deformation using a meshless method. In: Doyle, B., Miller, K., Wittek, A., Nielsen, P. M. F. (Eds.), *Computational biomechanics for medicine* (pp. 59–71). Springer.
3. Ciarlet, P. G. (1988). *Mathematical elasticity*. The Netherlands: North Hollad.
4. Dale, A. M., Fischl, B., & Sereno, M. I. (1999). Cortical surface-based analysis: I. Segmentation and surface reconstruction. *Neuroimage, 9*(2), 179–194.
5. Dora, L., Agrawal, S., Panda, R., & Abraham, A. (2017). State-of-the-art methods for brain tissue segmentation: A review. *IEEE Reviews in Biomedical Engineering, 10,* 235–249.
6. Fedorov, A., Beichel, R., Kalpathy Cramer, J., Finet, J., Fillion Robin, J.-C., Pujol, S., et al. (2012). 3D slicer as an image computing platform for the quantitative imaging network. *Magnetic Resonance Imaging, 30*(9), 1323–1341.
7. Garlapati, R. R., Aditi, R., Joldes, G. R., Wittek, A., Mostayed, A., Doyle, B., Warfield, S. K., Kikinis, R., Knuckey, N., Bunt, S., & Miller, K. (2013). Biomechanical modeling provides more accurate data for neuronavigation than rigid registration. *Journal of Neurosurgery.* Accepted for publication on 4th October, 2013.
8. Geuzaine, C., & Remacle, J. F. (2009). Gmsh: A 3-D finite element mesh generator with built-in pre-and post-processing facilities. *International Journal for Numerical Methods in Engineering, 79*(11), 1309–1331.
9. Horton, A., Wittek, A., Joldes, G. R., & Miller, K. (2010). A meshless total lagrangian explicit dynamics algorithm for surgical simulation. *International Journal for Numerical Methods in Biomedical Engineering, 26,* 977–998.
10. Jin, X., Joldes, G. R., Miller, K., Yang, K. H., & Wittek, A. (2014). Meshless algorithm for soft tissue cutting in surgical simulation. *Computer Methods in Biomechanics and Biomedical Engineering, 17,* 800–817.
11. Joldes, G., Bourantas, G., Zwick, B., Chowdhury, H., Wittek, A., Agrawal, S., et al. (2019). Suite of meshless algorithms for accurate computation of soft tissue deformation for surgical simulation. *Medical Image Analysis, 56,* 152–171.
12. Joldes, G. R., Wittek, A., & Miller, K. (2009). Suite of finite element algorithms for accurate computation of soft tissue deformation for surgical simulation. *Medical Image Analysis, 13*(6), 912–919.
13. Joldes, G. R., Wittek, A., & Miller, K. (2011). An adaptive dynamic relaxation method for solving nonlinear finite element problems. Application to brain shift estimation. *International Journal for Numerical Methods in Biomedical Engineering, 27*(2), 173–185.
14. Joldes, G. R., Wittek, A., Miller, K., & Morriss, L. (2008). Realistic and efficient brain-skull interaction model for brain shift computation. In: K. Miller and P. M. F. Nielsen, Computational Biomechanics for Medicine III Workshop, Miccai, pp. 95–105. New York.
15. Joldes, G. R., Wittek, A., Warfield, S. K., & Miller, K. (2012). Performing brain image warping using the deformation field predicted by a biomechanical model. In: *Computational Biomechanics for Medicine* (pp. 89–96). Springer.

16. Li, M., A. Wittek, G. R. Joldes and K. Miller (2016). Fuzzy Tissue Classification for Non-Linear Patient-Specific Biomechanical Models for Whole-Body Image Registration. Computational Biomechanics for Medicine: Imaging, Modeling and Computing. G. R. Joldes, B. Doyle, A. Wittek, P. M. F. Nielsen and K. Miller. Cham, Springer International Publishing: 85–96.
17. Lorensen, W. E. H. E. C. (1987). Marching cubes: A high resolution 3D surface construction algorithm. SIGGRAPH Comput. Graph. 21 (Association for Computing Machinery), pp. 163–169.
18. Miga, M. I., Sun, K., Chen, I., Clements, L. W., Pheiffer, T. S., Simpson, A. L., et al. (2016). Clinical evaluation of a model-updated image-guidance approach to brain shift compensation: experience in 16 cases. *International Journal of Computer Assisted Radiology and Surgery, 11*(8), 1467–1474.
19. Miller, K., Chinzei, K., Orssengo, G., & Bednarz, P. (2000). Mechanical properties of brain tissue in-vivo: Experiment and computer simulation. *Journal of Biomechanics, 33,* 1369–1376.
20. Miller, K., Horton, A., Joldes, G. R., & Wittek, A. (2012). Beyond finite elements: A comprehensive, patient-specific neurosurgical simulation utilizing a meshless method. *Journal of Biomechanics, 45*(15), 2698–2701.
21. Miller, K., & Lu, J. (2013). On the prospect of patient-specific biomechanics without patient-specific properties of tissues. *Journal of the Mechanical Behavior of Biomedical Materials, 27,* 154–166.
22. Miller, K., Wittek, A., & Joldes, G. (2011). *Biomechanical modeling of the brain for computer-assisted neurosurgery* (pp. 111–136). Springer, New York: Biomechanics of the Brain.
23. Mostayed, A., Garlapati, R., Joldes, G., Wittek, A., Roy, A., Kikinis, R., et al. (2013). Biomechanical model as a registration tool for image-guided neurosurgery: Evaluation against BSpline registration. *Annals of Biomedical Engineering, 41*(11), 2409–2425.
24. Neal, M. L., & Kerckhoffs, R. (2010). Current progress in patient-specific modeling. *Briefings in Bioinformatics, 11,* 15.
25. Otsu, N. (1979). A threshold selection method from gray-level histograms. *IEEE Transactions on Systems, Man, and Cybernetics, 9*(1), 62–66.
26. Pruthi, S., Dawant, B., & Parker, S. L. Initial experience with using a structured light 3D scanner and image registration to plan bedside subdural evacuating port system placement.
27. Valette, S., Chassery, J. M., & Prost, R. (2008). Generic remeshing of 3D triangular meshes with metric-dependent discrete Voronoi diagrams. *IEEE Transactions on Visualization and Computer Graphics, 14*(2), 369–381.
28. Wittek, A., Grosland, N., Joldes, G., Magnotta, V., & Miller, K. (2016). From finite element meshes to clouds of points: A review of methods for generation of computational biomechanics models for patient-specific applications. *Annals of Biomedical Engineering, 44*(1), 3–15.
29. Wittek, A., Hawkins, T., & Miller, K. (2009). On the unimportance of constitutive models in computing brain deformation for image-guided surgery. *Biomechanics and Modeling in Mechanobiology, 8,* 77–84.
30. Wittek, A., Joldes, G., Couton, M., Warfield, S. K., & Miller, K. (2010). Patient-specific non-linear finite element modelling for predicting soft organ deformation in real-time; Application to non-rigid neuroimage registration. *Progress in Biophysics and Molecular Biology, 103,* 292–303.
31. Wittek, A., Miller, K., Kikinis, R., & Warfield, S. K. (2007). Patient-specific model of brain deformation: Application to medical image registration. *Journal of Biomechanics, 40,* 919–929.
32. Zhang, J. Y., Joldes, G. R., Wittek, A., & Miller, K. (2013). Patient-specific computational biomechanics of the brain without segmentation and meshing. *International Journal for Numerical Methods in Biomedical Engineering, 29*(2), 293–308.
33. Zhang, Y. J., Joldes, G. R., Wittek, A., & Miller, K. (2013). Patient-specific computational biomechanics of the brain without segmentation and meshing. *International Journal for Numerical Methods in Biomedical Engineering, 29*(2), 16.

Computer Simulation of the Resection Induced Brain Shift; Preliminary Results

Yue Yu, George Bourantas, Tina Kapur, Sarah Frisken, Ron Kikinis, Arya Nabavi, Alexandra Golby, Adam Wittek, and Karol Miller

Abstract Neurosurgery for tumour resection requires precise planning and navigation to maximise tumour removal and at the same time protect healthy tissue. Biomechanical modelling and computer simulation provide a means for brain deformation prediction, facilitating millimetre-accurate neuronavigation. In this contribution, we modelled brain deformation due to tumour resection. We assumed that the brain deformations were caused by stresses on the resected tumour surface released when the tumour is removed. We computed deformations using nonlinear finite element analysis in Abaqus FEA software suite. We used the patient-specific geometry extracted from the pre-operative magnetic resonance image (MRI). We included the parenchyma tissue, cerebral tumour, cerebral ventricles, skull and cerebral falx in our computational biomechanics brain model. We created mixed (consisting of hexahedral and tetrahedral elements) patient-specific finite element mesh, and used realistic material properties together with appropriate contact conditions at boundaries. We compared our predicted intra-operative brain geometry (based on the calculated deformation field) contours with the intra-operative MRI to assess the accuracy of our results. The results indicate that our biomechanical models can complement medical image processing techniques when conducting non-rigid registration.

Keywords Brain shift · Biomechanics · Finite element method · Tumour resection · Image-guided surgery

Y. Yu (✉) · G. Bourantas · A. Wittek · K. Miller
Intelligent systems for Medicine Laboratory, Perth, Western Australia, Australia
e-mail: yue.yu@research.uwa.edu.au

T. Kapur · S. Frisken · R. Kikinis · A. Golby
Brigham and Women's Hospital, Harvard Medical School, Boston, MA, USA

A. Nabavi
Department of Neurosurgery, KRH, Klinikum Nordstadt, Hannover, Germany

K. Miller
Harvard Medical School, Boston, MA, USA

© The Author(s), under exclusive license to Springer Nature Switzerland AG 2021
K. Miller et al. (eds.), *Computational Biomechanics for Medicine*,
https://doi.org/10.1007/978-3-030-70123-9_2

17

1   Introduction

Accurate delineation of the brain tumour boundaries and assessment of the structural and functional anatomy are crucially important to minimise the risk of neurological deficits due to surgery. High quality pre-operative MRI provides details of the patient's brain anatomy. However, the movements of the brain ("brain shift") during the surgery distort the pre-operative anatomy and complicate the complete removal of the brain tumour. Intra-operative MRI provides accurate imaging guidance and help to track the changes of the patient's brain anatomy induced by surgery [4]. Nonetheless, it is a troublesome equipment to use [24]. It is also very expensive and not commonly available [22]. The initial investment for an intra-operative MRI suite is between 5 to 10 million dollars [25].

Biomechanical modelling has been extensively used to calculate the brain shift and to register a high-quality pre-operative MRI onto an intra-operative configuration of a patient [20]. Such registration can be achieved without intra-operative MRI [20]. Most of the studies of brain deformation estimation based on biomechanical model have focused on the early surgical stages (prior to the surgical dissection, tumour removal). Biomechanical modelling and computer simulation provide rapid and accurate registration for craniotomy-induced brain shift [5, 27, 28]. In this contribution, we extend this methodology to the estimation of the brain shift due to tumour resection.

Brain shift is dependent upon various factors such as cerebrospinal fluid (CSF) drainage, gravity, surgical tools, resection, and swelling [6, 21]. In this study, we are taking into account the effect of gravity and the decompression force along the boundary of the tumour cavity. We estimated the brain shift induced by tumour resection, using the commercial finite element software Abaqus FEA [10]. The biomechanical brain model in this study is created based on a real clinical case.

This paper is organised as follows: In Sect. 2, we presented the methodology of this study. In Sect. 3, we showed our results and its assessment. Section 4 contains discussion and conclusions.

2   Methods

2.1   Biomechanical Modelling Strategy

The patient underwent a resection of left parietal metastatic poorly differentiated carcinoma with focal neuroendocrine features at the Department of Surgery, Brigham and Women's Hospital (Harvard Medical School, Boston, MA, USA).

We assumed that before opening of the skull, buoyancy force due to the intracranial fluids balances the gravity, so that the patient's brain is in an unloaded state pre-operatively. Opening of the skull disrupted such balance due to the pressure release and the drainage of CSF. The brain deformed under the effect of the gravity and stresses developed in response to gravity loading. We then applied the internal forces released after tumour was removed to the remaining brain tissue, and solved for displacements.

2.2   Construction of the Finite Element Mesh

We aligned the high quality pre-operative MRI onto the intra-operative configuration using rigid registration. The pre-operative image was segmented using 3D Slicer image computing platform [2] to obtain the anatomical features of interest. We used the segmentation algorithms available in Segment Editor module in 3D Slicer: threshold effects [23] followed by manual correction. The tumour segmentation in this study includes the gadolinium-enhanced areas. We created a three-dimensional surface model (Fig. 1) with three distinctive features (parenchyma tissues, ventricles, cerebral tumour) of the specific patient's brain based on the pre-operative MRI segmentations.

We used first order elements (linear tetrahedron and linear hexahedron) in the biomechanical brain model to facilitate efficient computation. We applied IA-FEMesh open-source meshing software toolkit [7] to generate the hexahedral elements. Hexahedral elements with Jacobian less than 0.3 [15], and internal angles less than 45 degrees or larger than 135 degrees are regarded as poor-quality [1]. We replaced poor-quality elements with linear (4-noded) tetrahedrons using HyperMesh (commercial finite element mesh generator by Altair of Troy, MI, USA). We used

Fig. 1 Patient-specific brain surface extracted from the pre-operative MRI used in this study. The ventricles are in blue, tumour is in red, and parenchyma tissue is in grey

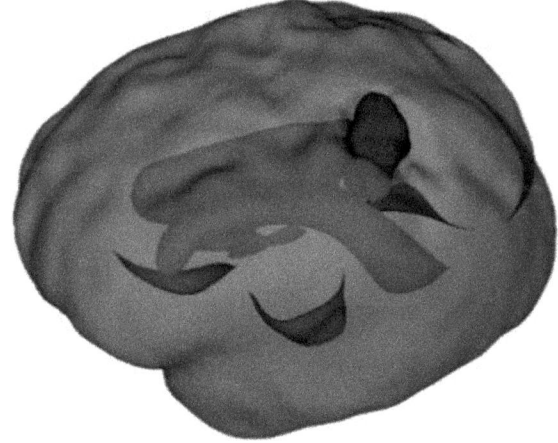

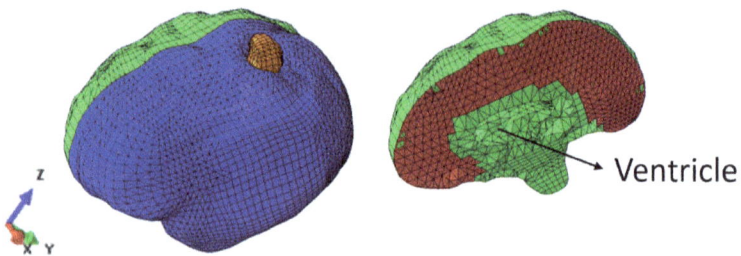

Fig. 2 Patient-specific finite element mesh of the brain constructed here based on the real clinical case. The brain mesh contains parenchyma tissues, cerebral tumour, cerebral falx, and ventricle (presented by hollow space)

hybrid elements (Abaqus FEA element types C3D4H and C3D8H) to avoid the volumetric locking [10]. The entire volumetric mesh comprises of 19,811 nodes, 40,353 elements, resulting on the model with ~ 60,000 degrees of freedom.

Figure 2 shows the finite element mesh of the brain created in this study. Blue mesh represents the left brain hemisphere, green mesh represents the right brain hemisphere, and orange mesh represents the tumour. Left and right brain hemispheres are separated by the cerebral falx. Red 2D triangle mesh represents the cerebral falx, and the hollow spaces inside the parenchyma tissue are the brain ventricles.

2.3 Materials Properties

Parenchyma tissue is considered as an almost incompressible material and, in most literature, the Poisson's ratio is between 0.45 to 0.49 [17]. In the analysed brain, the left hemisphere presented clear signs of atrophy, resulting in increased fluid content and decreased cellular tissue content. We accounted for this anomaly by using the lower Poisson's ratio = 0.45 that allows slight compressibility of this left brain hemisphere and accounts for the drainage of CSF after skull opening. The Poisson's ratio is 0.49 for the right brain hemisphere. We used Ogden constitutive model to describe constitute properties of the brain. Ogden model accounts well for the brain tissue deformation and we assigned 842 Pa and -4.7 to the shear modulus μ_0 and material constant α respectively [18].

In Abaqus FEA [10] the following formulation is used to express the Ogden strain energy potential,

$$U = \frac{2\mu_0}{\alpha^2} \left(\bar{\lambda}_1^\alpha + \bar{\lambda}_2^\alpha + \bar{\lambda}_3^\alpha - 3 \right) + \frac{\kappa_0}{2}(J - 1)^2, \tag{1}$$

where $\bar{\lambda}_i$ are the deviatoric principle stretches, μ_0 is the instantaneous shear modulus, α is a material constant, J is the determinant of the deformation gradient, and κ_0 is the bulk modulus.

Table 1 Material properties of biomechanical model

Model components	Density (kg/m^3)	Shear modulus (μ_0) (Pa)	Material constant (α)	Poisson's ratio
Parenchyma (left)	1000	842 [18]	−4.7 [18]	0.45
Parenchyma (right)	1000	842 [18]	−4.7 [18]	0.49
Tumour	1000	2526 [26]	−4.7 [26]	0.49
Ventricle	Modelled as hollow space			
Falx	Rigid			
Skull	Rigid			

For the brain tumour, we used the same constitutive model as the parenchyma tissues, and its shear modulus was assigned a value three times larger than that of the parenchyma tissue [26] (Table 1).

2.4 Boundary Conditions

The stiffness of skull is several orders of magnitude higher than that of the brain tissue. Therefore, the skull is modelled as a rigid body. To define the boundary conditions between brain surface and skull, a frictionless finite sliding contact is defined between brain and skull [13]. Such boundary conditions have been widely used in computing the brain deformation in our previous research [20, 28].

The cerebral falx is an infolding of the dura matter that is rigidly attached to the top of the skull along the sagittal midline. It divides the brain into the left and right brain hemispheres. The falx is also surrounded by the CSF along the lateral faces and inferior aspect just above the corpus callosum. The contact interactions between the falx and brain hemispheres were modelled as a frictionless contact, which allows the cerebral hemispheres to freely slide tangentially [9, 19]

As the brain stem connects to the spinal cord, we chose to fix the lower part of the brain stem.

2.5 Loading

The gravity is applied to the model in the Y direction of the coordinate system in the intra-operative configuration. Traction forces are applied on the nodes (Fig. 3c) after the tumour is removed from the model.

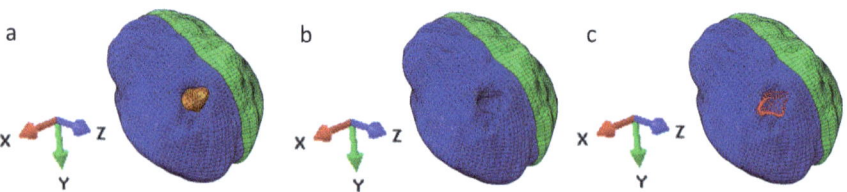

Fig. 3 Patient-specific brain model created and used in this study in the intra-operative config-uration. **a** Complete biomechanical brain model; **b** Biomechanical brain model without tumour. **c** Nodes on the tumour surface, shared between the tumour and healthy parenchyma (where traction forces are calculated and applied), showed as red dots. Gravity acts in Y direction

2.6 Computer Simulation

We used commercial finite element software Abaqus FEA non-linear static solver [10]. We divided the simulation into two stages. In Stage 1, we computed the defor-mation field resulting from the gravity. We obtained the internal forces of the nodes on the tumour surface, shared between the tumour and healthy parenchyma. In Stage 2, we simulated the tumour resection-induced brain shift. We removed the tumour from the mesh and applied the nodal reaction forces computed in Stage 1 to the rest of the model. As from the visual analysis of the intra-operative images, we concluded that the tumour resection was complete, all finite elements discretising the tumour were removed.

2.7 Image Warping

We obtained the deformation field from the biomechanical model. Then we used it to warp the pre-operative image, so that it corresponds to the intra-operative configu-ration of the brain. Mesh nodal coordinates are extracted and exported from Abaqus FEA to 3D Slicer. We obtained the BSpline transform using 'Scattered Transform' module in 3D Slicer. 'Scattered Transform' interpolates displacements at nodes using a BSpline algorithm [14]. Once the BSpline transformation matrix is obtained, we used this transformation matrix to warp the pre-operative MRI. The predicted intra-operative geometry contour showed in Result section is extracted from the warped image.

3 Results

Figure 4a shows the initial (undeformed) state of the brain mesh; Fig. 4b shows the brain mesh deformed under the gravity; Fig. 4c shows the equilibrium state of the brain after the tumour is removed. The simulation predicted that the maximum displacement (8.4 mm) occurred in the cortical surface of the left frontal lobe.

Figure 5 shows the comparison of the undeformed and deformed brain model in axial, sagittal, and coronal slices/sections. Slices for the undeformed (pre-operative) brain model are in grey. The computed deformations are indicated by the colour scale in the deformed shape slices (Figs. 6 and 7).

In image-guided surgery, brain shift estimation accuracy is typically assessed by evaluating the accuracy of agreement between the registered position of the pre-operative image and the real patient position usually established by an intra-operative MRI [27]. Validation of the biomechanical model predictions has been often done by tracking the manually selected landmarks in MRIs [3]. However, determining

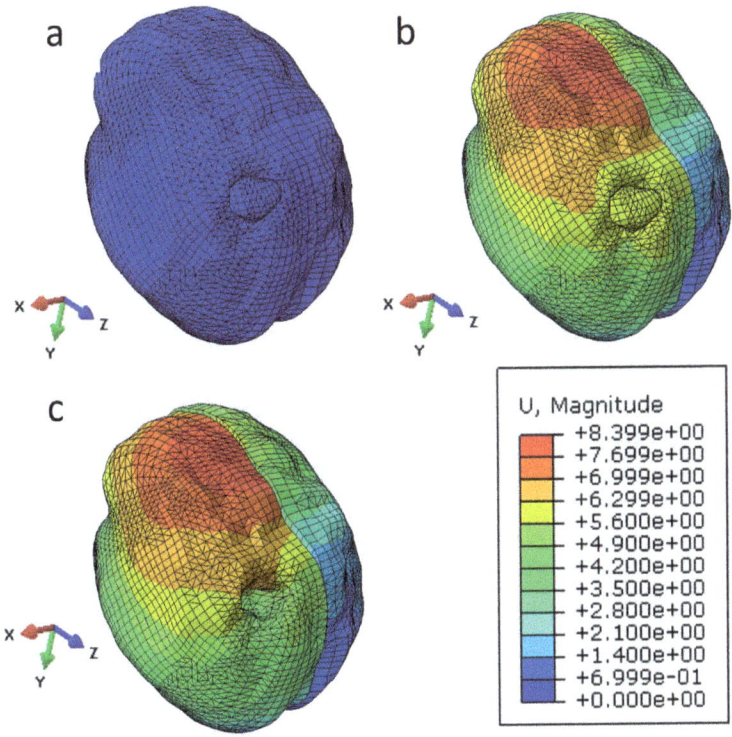

Fig. 4 Displacement magnitude (in mm) predicted using our brain model in colour scale. **a** undeformed brain model. **b** deformation resulting from the application of gravity. **c** deformed brain model after the tumour removal

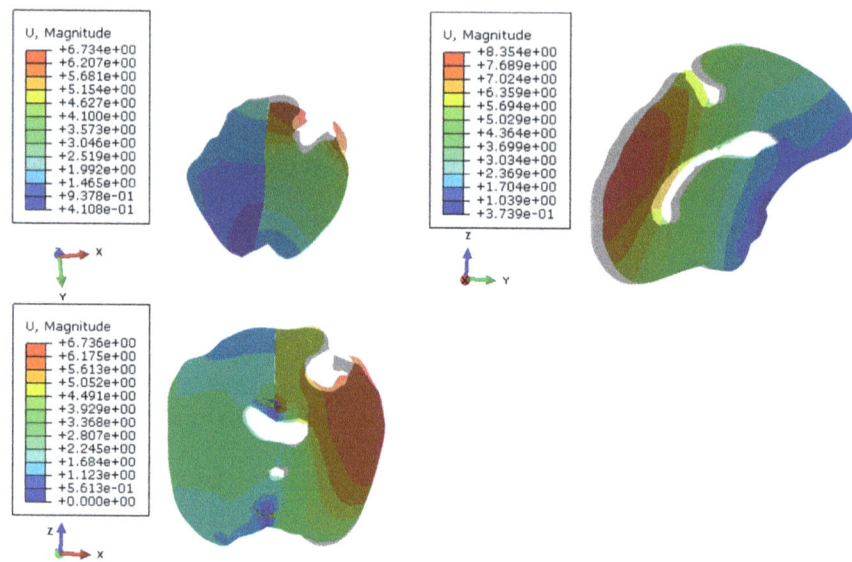

Fig. 5 Slices/sections through the undeformed and deformed biomechanical brain model. Slices for the undeformed (pre-operative) brain model are in grey. The deformations (in mm) are indicated by the colour scale in the deformed shape slices. Location of the cross-section planes is shown in Fig. 6

these landmarks locations between MRIs is very time-consuming and its accuracy is questionable relying on the experience of the trained neurosurgeon [16].

We provide qualitative estimation of our results by showing the intra-operative MRI (acquired after tumour was resected) overlaid with the contours of predicted intra-operative brain and ventricle (warped using the predicted deformation field) contours (Fig. 8) and its corresponding cross sections (Fig. 6). The Contours were obtained using Segment Editor (threshold effects [23] with manual correction) in 3D Slicer. The comparison showed in Fig. 8 indicates good alignment between the predicted intra-operative contours and intra-operative MRI.

Segmenting ventricles on the intraoperative image would introduce error that is difficult to quantify. Therefore, we provide only approximate quantitative error measurement. The maximum difference between the predicted and intra-operative ventricle contours is 8.6 mm. This difference is indicated in Fig. 7 by the error scale.

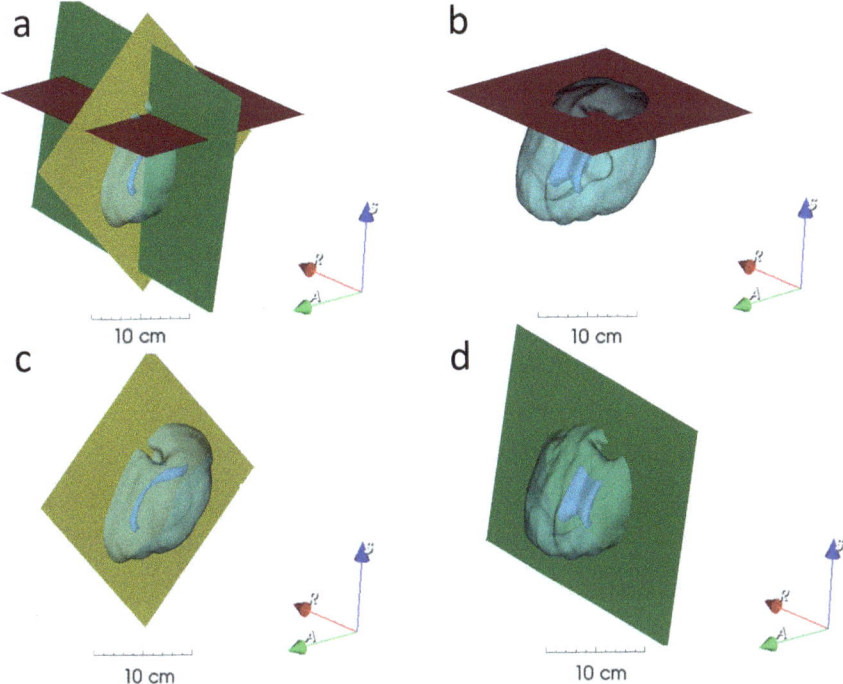

Fig. 6 Location of the planes for sections showed in Figs. 5 and 8. Red represents the axial section, yellow—sagittal section, and green - coronal section. The coordinate is RAS (Right-Anterior-Superior) system

4   Discussion and Conclusions

Brain shift is a continuous process that evolves differently in various distinct brain regions and the estimation of the brain shift induced by the tumour resection is a difficult task. The results of this study suggest that the application of our biomechanics-based approach could estimate the resection-induced brain shift sufficiently accurately to be potentially useful in neuronavigation and planning. Another significant advantage of using the biomechanical model is that it does not require the intra-operative MRI at all to compute the deformation. Only the pre-operative MRI and the volume of tissue removed is needed.

However, limitations of our simulations exist. The establishment of the biomechanical mesh is a laborious process and the accuracy of the brain model highly depends on the segmentation of the pre-operative neuroimages. Simulation of the tumour resection-induced brain shift involves large displacement and large strain, with element distortion negatively affecting the accuracy of the finite element results. Meshless methods that have ability to deal with extremely large deformations and boundary changes such as the changes induced by tumour resection [11, [12] may be more appropriate here. The fluid content in the left hemisphere of the analysed

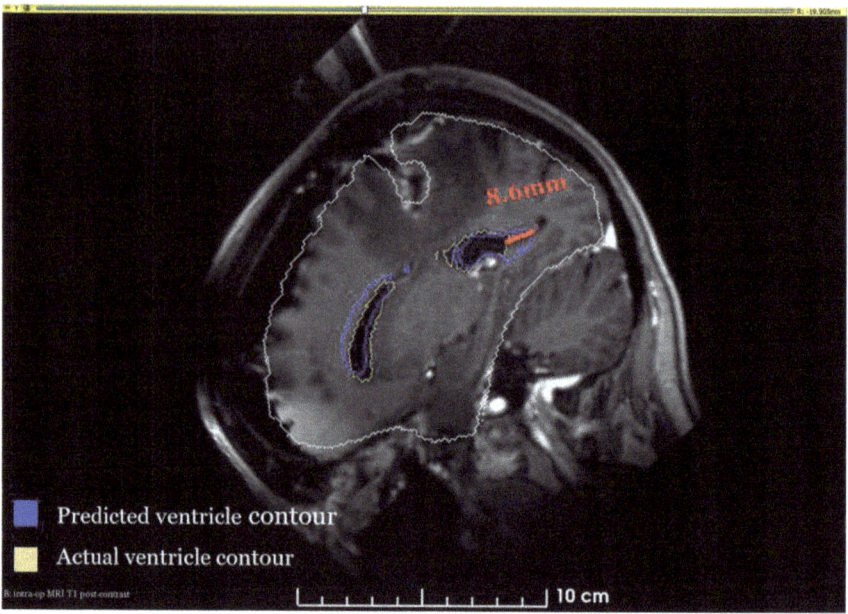

Fig. 7 Maximum difference (red line) occurs between the predicted (blue) and intra-operative (yellow) ventricle contours

patient case would be extremely difficult to discretise using a finite element mesh. The fussy tissue classification algorithm, which assigns the material properties through a cloud of nodes [29], may provide solution to this challenge. Furthermore, appropriate intra-operative information could increase the accuracy of the biomechanical based simulation [8].

The presented results of application of patient-specific biomechanics model to predict the intra-operative brain deformation induced by the tumour resection are very promising. However, as only one tumour resection procedure was analysed, they can be regarded as a proof of concept rather than complete evaluation of the accuracy of our methods. Such evaluation would require an extensive study on a large cohort of patients. Therefore, in our future work, we intend to move from the finite element solver to our inhouse Meshless Total Lagrangian Explicit Dynamics (MTLED) algorithm [11, 12]. This will facilitate creating more comprehensive biomechanical models with a shorter pre-processing time and obtaining more robust results.

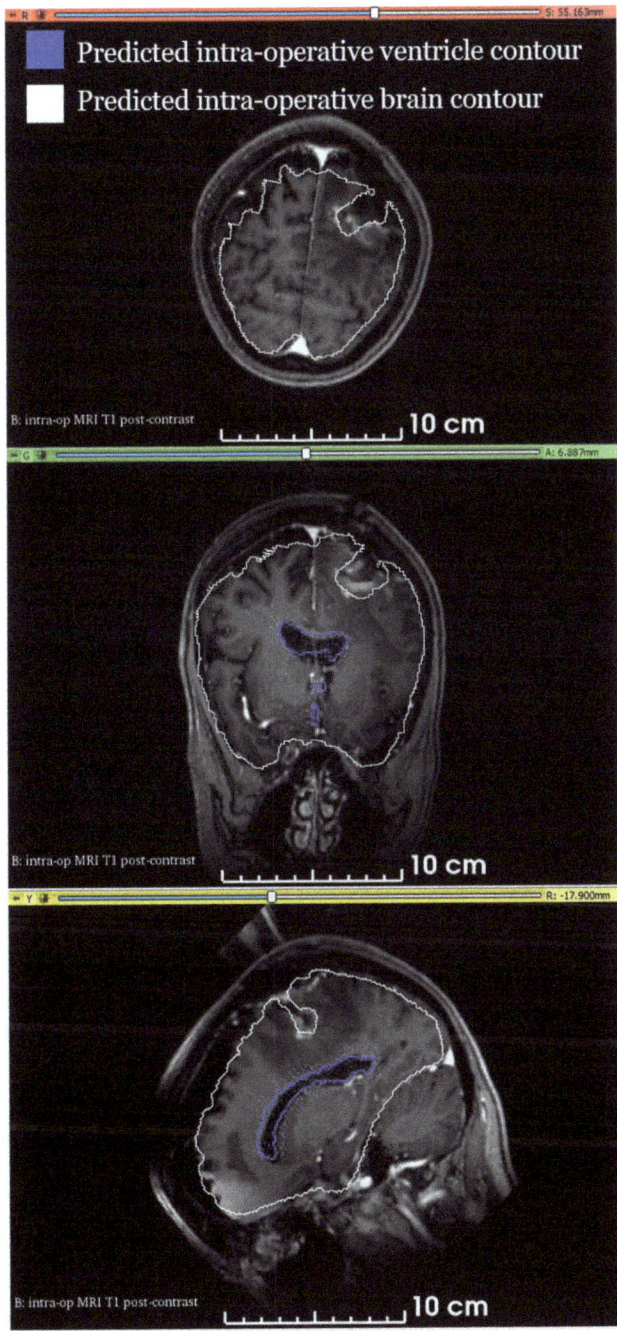

Fig. 8 Comparison of intra-operative MRI with the predicted (using our patient-specific computational biomechanics model) intra-operative contours of the brain ventricles (blue) and cortical
surface (white). Resolution of the intra-operative MRI in this study is 1.0 mm × 1.0 mm × 1.2 mm

Acknowledgements The first author is a recipient of the Research Training Program (RTP) scholarship and acknowledges the financial support of the University of Western Australia. Funding from National Health and Medical Research Council (NHMRC) project grant APP1144519 is gratefully acknowledged.

References

1. Couteau, B., Payan, Y., & Lavallée, S. (2000). The mesh-matching algorithm: An automatic 3D mesh generator for finite element structures. *Journal of Biomechanics, 33*(8), 1005–1009. https://doi.org/10.1016/S0021-9290(00)00055-5.
2. Fedorov, A., Beichel, R., Kalpathy-Cramer, J., Finet, J., Fillion-Robin, J.-C., Pujol, S., et al. (2012). 3D Slicer as an image computing platform for the quantitative imaging network. *Magnetic Resonance Imaging, 30*(9), 1323–1341. https://doi.org/10.1016/j.mri.2012.05.001.
3. Ferrant, M., Nabavi, A., Macq, B. t., Black, P. M., Jolesz, F. A., Kikinis, R., & Warfield, S. K. (2002). Serial registration of intraoperative MR images of the brain. *Medical Image Analysis, 6*(4), 337–359. 10.1016/S1361-8415(02)00060-9.
4. Gandhe, R., & Bhave, C. (2018). Intraoperative magnetic resonance imaging for neurosurgery—An anaesthesiologist's challenge. *Indian Journal of Anaesthesia, 62*(6), 411–417. https://doi.org/10.4103/ija.IJA_29_18.
5. Garlapati, R. R., Roy, A., Joldes, G. R., Wittek, A., Mostayed, A., Doyle, B., et al. (2014). More accurate neuronavigation data provided by biomechanical modeling instead of rigid registration. *Journal of Neurosurgery, 120*(6), 1477–1483. https://doi.org/10.3171/2013.12. JNS131165.
6. Gerard, I. J., Kersten-Oertel, M., Petrecca, K., Sirhan, D., Hall, J. A., & Collins, D. L. (2017). Brain shift in neuronavigation of brain tumors: A review. *Medical Image Analysis, 35*(C), 403–420. https://doi.org/10.1016/j.media.2016.08.007.
7. Grosland, N. M., Shivanna, K. H., Magnotta, V. A., Kallemeyn, N. A., DeVries, N. A., Tadepalli, S. C., et al. (2008). IA-FEMesh: An open-source, interactive, multiblock approach to anatomic finite element model development. *Computer Methods and Programs in Biomedicine, 94*(1), 96–107. https://doi.org/10.1016/j.cmpb.2008.12.003.
8. Hajnal, J. V., Hawkes, D. J., & Hill, D. L. G. (2001). *Medical image registration.* Boca Raton, FL: CRC Press.
9. Hernandez, F., Giordano, C., Goubran, M., Parivash, S., Grant, G., Zeineh, M., et al. (2019). Lateral impacts correlate with falx cerebri displacement and corpus callosum trauma in sports-related concussions. *Biomechanics and Modeling in Mechanobiology, 18*(3), 631–649. https://doi.org/10.1007/s10237-018-01106-0.
10. Hibbitt, Karlsson, & Sorensen. (2014). *ABAQUS/Standard User's Manual, Version 6.14*: Simulia, Providence, RI.
11. Horton, A., Wittek, A., Joldes, G. R., & Miller, K. (2010). A meshless total Lagrangian explicit dynamics algorithm for surgical simulation. *International Journal for Numerical Methods in Biomedical Engineering, 26*(8), 977–998. https://doi.org/10.1002/cnm.1374.
12. Joldes, G., Bourantas, G., Zwick, B., Chowdhury, H., Wittek, A., Agrawal, S., et al. (2019). Suite of meshless algorithms for accurate computation of soft tissue deformation for surgical simulation. *Medical Image Analysis, 56*, 152–171. https://doi.org/10.1016/j.media. 2019.06.004.
13. Joldes, G., Wittek, A., Miller, K., & Morriss, L. (2008). Realistic and efficient brain-skull interaction model for brain shift. *Computation, 1*, 1–12.
14. Joldes, G. R., Wittek, A., Warfield, S. K., & Miller, K. (2012). Performing brain image warping using the deformation field predicted by a biomechanical model. In (2012 ed., pp. 89–96). New York, NY: Springer.

15. Li, Z., Hu, J., Reed, M. P., Rupp, J. D., Hoff, C. N., Zhang, J., et al. (2011). Development, validation, and application of a parametric pediatric head finite element model for impact simulations. *Annals of Biomedical Engineering, 39*(12), 2984–2997. https://doi.org/10.1007/s10439-011-0409-z.

16. Miga, M. I., Paulsen, K. D., Lemery, J. M., Eisner, S. D., Hartov, A., Kennedy, F. E., et al. (1999). Model-updated image guidance: Initial clinical experiences with gravity-induced brain deformation. *IEEE Transactions on Medical Imaging, 18*(10), 866–874. https://doi.org/10.1109/42.811265.

17. Miller, K. (2019). *Biomechanics of the brain* (2nd ed.). Cham, Switzerland: Springer.

18. Miller, K., & Chinzei, K. (2002). Mechanical properties of brain tissue in tension. *Journal of Biomechanics, 35*(4), 483–490. https://doi.org/10.1016/S0021-9290(01)00234-2.

19. Morin, F., Courtecuisse, H., Reinertsen, I., Le Lann, F., Palombi, O., Payan, Y., et al. (2017). Brain-shift compensation using intraoperative ultrasound and constraint-based biomechanical simulation. *Medical Image Analysis, 40,* 133–153. https://doi.org/10.1016/j.media.2017.06.003.

20. Mostayed, A., Garlapati, R. R., Joldes, G. R., Wittek, A., Roy, A., Kikinis, R., et al. (2013). Biomechanical model as a registration tool for image-guided neurosurgery: Evaluation against BSpline registration. *Annals of Biomedical Engineering, 41*(11), 2409–2425. https://doi.org/10.1007/s10439-013-0838-y.

21. Nabavi, T. A., McL. Black, S. P., Gering, K. D., Westin, B. C.-F., Mehta, M. V., Pergolizzi, A. R., … Jolesz, A. F. (2001). Serial intraoperative magnetic resonance imaging of brain shift. *Neurosurgery, 48*(4), 787–798. 10.1097/00006123-200104000-00019.

22. Pichierri, A., Bradley, M., & Iyer, V. (2019). Intraoperative magnetic resonance imaging-guided glioma resections in awake or asleep settings and feasibility in the context of a public health system. *World Neurosurgery: X, 3.* https://doi.org/10.1016/j.wnsx.2019.100022.

23. Pinter, C. (2017). *Segmentation for 3D printing.* Retrieved from Canada: https://github.com/SlicerRt/SlicerRtDoc/raw/master/tutorials/SegmentationFor3DPrinting_TutorialContestWinter2017.pdf.

24. Reddy, U., White, M. J., & Wilson, S. R. (2012). Anaesthesia for magnetic resonance imaging. *Continuing Education in Anaesthesia Critical Care & Pain, 12*(3), 140–144. https://doi.org/10.1093/bjaceaccp/mks002.

25. Shah, M. N., Leonard, J. R., Inder, G. E., Gao, F., Geske, M., Haydon, D. H., et al. (2012). Intraoperative magnetic resonance imaging to reduce the rate of early reoperation for lesion resection in pediatric neurosurgery: Clinical article. *Journal of Neurosurgery: Pediatrics, 9*(3), 259–264. https://doi.org/10.3171/2011.12.PEDS11227.

26. Wittek, A., Hawkins, T., & Miller, K. (2009). On the unimportance of constitutive models in computing brain deformation for image-guided surgery. *Biomechanics and Modeling in Mechanobiology, 8*(1), 77–84. https://doi.org/10.1007/s10237-008-0118-1.

27. Wittek, A., Joldes, G., Couton, M., Warfield, S. K., & Miller, K. (2010). Patient-specific non-linear finite element modelling for predicting soft organ deformation in real-time; Application to non-rigid neuroimage registration. *Progress in Biophysics and Molecular Biology, 103*(2), 292–303. https://doi.org/10.1016/j.pbiomolbio.2010.09.001.

28. Wittek, A., Miller, K., Kikinis, R., & Warfield, S. K. (2006). Patient-specific model of brain deformation: Application to medical image registration. *Journal of Biomechanics, 40*(4), 919–929. https://doi.org/10.1016/j.jbiomech.2006.02.021.

29. Zhang, J. Y., Joldes, G. R., Wittek, A., & Miller, K. (2013). Patient-specific computational biomechanics of the brain without segmentation and meshing. *International Journal for Numerical Methods in Biomedical Engineering, 29*(2), 293–308. https://doi.org/10.1002/cnm.2507.

Mandibular Teeth Movement Variations in Tipping Scenario: A Finite Element Study on Several Patients

Torkan Gholamalizadeh, Sune Darkner, Paolo Maria Cattaneo, Peter Søndergaard, and Kenny Erleben

Abstract Previous studies on computational modeling of tooth movement in orthodontic treatments are limited to a single model and fail in generalizing the simulation results to other patients. To this end, we consider multiple patients and focus on tooth movement variations under the identical load and boundary conditions both for intra- and inter-patient analyses. We introduce a novel computational analysis tool based on finite element models (FEMs) addressing how to assess initial tooth displacement in the mandibular dentition across different patients for uncontrolled tipping scenarios with different load magnitudes applied to the mandibular dentition. This is done by modeling the movement of each patient's tooth as a nonlinear function of both load and tooth size. As the size of tooth can affect the resulting tooth displacement, a combination of two clinical biomarkers obtained from the tooth anatomy, i.e., crown height and root volume, is considered to make the proposed model generalizable to different patients and teeth.

Keywords Tooth movement modeling · Tooth movement variations · Finite element model · Computational analysis · Periodontal ligament · Biomechanical model

T. Gholamalizadeh (✉) · P. Søndergaard
3Shape A/S, Copenhagen, Denmark
e-mail: torkan@di.ku.dk

P. Søndergaard
e-mail: peter.soendergaard@3shape.com

T. Gholamalizadeh · S. Darkner · K. Erleben
Department of Computer Science, University of Copenhagen, Copenhagen, Denmark
e-mail: darkner@di.ku.dk

K. Erleben
e-mail: kenny@di.ku.dk

P. M. Cattaneo
Department of Dentistry and Oral Health, Aarhus University, Aarhus, Denmark
e-mail: paolo.cattaneo@dent.au.dk

31

1 Introduction

Orthodontic tooth movement is the result of alveolar bone remodeling caused by the applied forces and deformations in the periodontium. Finite element models (FEMs) is widely used to assess stress/strain in the alveolar bone and periodontal ligament (PDL), the fibrous connective tissue between tooth and bone, in the orthodontic treatments [5, 7, 13, 17, 20, 29]. Moreover, the initial and long-term tooth movements can be investigated using these models. In this work, we use an FEM to provide a biomechanical model of the full mandibular dentition focusing on initial teeth displacements caused by the applied load on the teeth (see Figs. 1 and 2). We generate patient-specific FEMs for three patients by segmenting the cone-beam computed tomography (CBCT) scans of the patients. The models are provided by using the same boundary conditions under the same scenarios. In each scenario, an identical force magnitude is applied perpendicular to the surface of each tooth to mimic an uncontrolled tipping movement. Besides, the load magnitude can change from 0.3 N

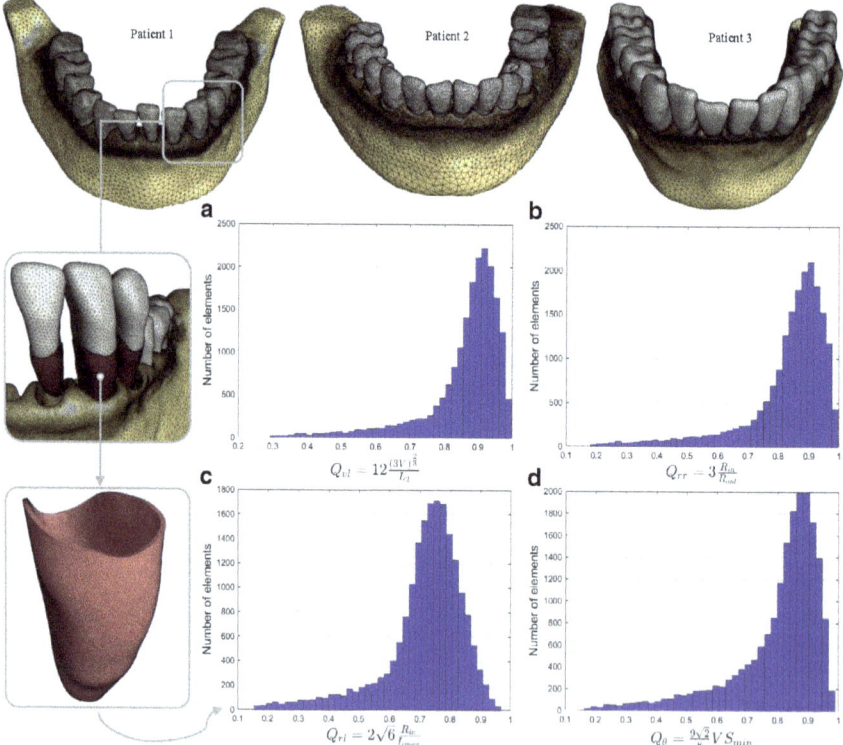

Fig. 1 FEMs and meshes for three patients, and mesh quality histograms for the PDL of the left canine. **a**: The volume-edge ratio, **b**: The radius ratio, **c**: The radius-edge ratio, **d**: A metric introduced in [12]

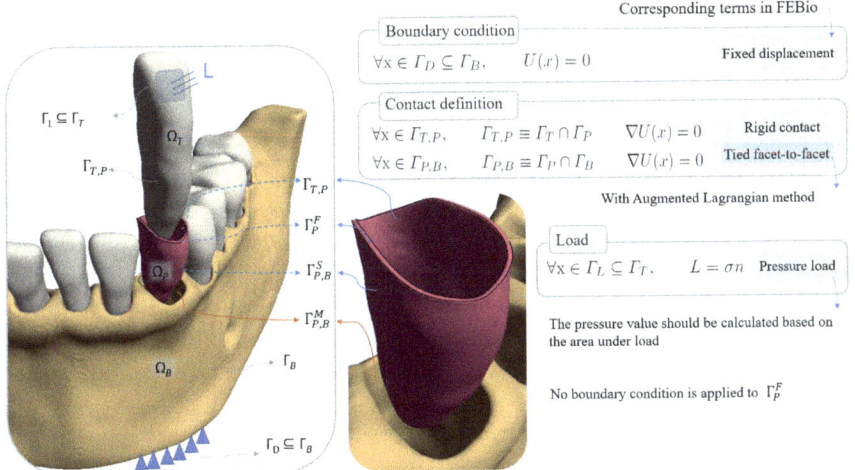

Fig. 2 A closeup view of the tooth supporting complex. The contacts between different domains and boundary conditions are presented. Tooth (Ω_T), Periodontal ligament (Ω_P), Alveolar bone (Ω_B), Tooth-PDL contact ($\Gamma_{T,P}$), PDL-Bone contact ($\Gamma_{P,B}$) and a Dirichlet boundary condition (Γ_D) are shown

to 1 N with 0.1 N increments, and teeth transformations are recorded for all teeth of each patient. Finally, the results are compared with the corresponding teeth of other patients.

Our hypothesis is that variations in the teeth anatomy of different patients and the load magnitudes can affect the resulting tooth displacement. Therefore, in this study, tooth movements (i.e., rotation and translation) are estimated as nonlinear functions of both load and the ratio of crown height to root volume using the obtained biomechanical models.

2   Related Work

The performance of the orthodontic treatments can be improved if the movement of the teeth could be predicted in a reliable way. Therefore, many studies have focused on predicting tooth movements in orthodontic treatments using FEMs. In general, the tooth movement occurs in two phases [13]. In the first phase [5, 6, 19, 29], which is the main focus of this study, tooth moves within the PDL space in few seconds after applying a force [18]. This movement is substantially due to the deformation of the PDL tissue caused by the applied load. In the second phase [7, 13, 17, 20], the resulting stress in the PDL and bone tissue causes a bone remodeling process, where the bone is resorbed and formed in the compressed and stretched regions of the PDL, respectively.

In the context of FE-based modeling of the initial tooth movement, some studies [6, 29] have investigated different types of movements individually including bodily movement, controlled tipping, and uncontrolled tipping. Some others have explored the teeth mesialization, distalization, or retraction scenarios [17, 19, 20, 27]. These studies have considered the effect of the force direction [17, 29], moment-to-force [6, 29], and force magnitude [6, 24] on tooth transformation [6, 17, 27] or location of the center of rotation [6, 29]. However, the jaw model, force system, and number of teeth used in the analyses are not consistent. For example, [6, 29] used a small portion of jaw, while [15] worked on a fully segmented jaw model. Likewise, different studies have examined different number of teeth, e.g., using a single tooth [29], two [6, 19] or more [17, 27]. The force and/or moments have also been applied to different parts including the surface of tooth [6, 19], center of the resistance [29], and orthodontic appliances [17, 20].

The abovementioned biomechanical models, however, might not be applicable for analyzing different teeth motions obtained from multiple patients. In other words, the obtained tooth displacement results represented only by visualizing the displacement fields [19, 27], measuring the displacement of the selected landmarks [6, 27], or acquiring the translations/rotations using some predefined measurement points [17] lack useful information about different tooth motion tendencies for full dentition of multiple patients and, hence, are less interpretable when it comes to the across patients modeling analyses.

Moreover, existing FEMs applied in computational orthodontics are mostly limited to a single patient [6, 15, 17, 19, 27, 29]. Although Likitmongkolsakul et al. [20] propose a stress-movement function of a canine for two orthodontic patients under an identical scenario, to the best of our knowledge, there are no other studies considering multiple patients for tooth movement modeling.

In this work, by considering the biomechanical models of human mandible acquired from CBCT scans of three patients, we investigate the tooth movement variations in multiple patients using rigid body transformations under different load magnitudes. To the best of our knowledge, this is the first computational model in orthodontics applied to three different patients. Our experiments consist of both intra- and inter-patient analyses. Considering teeth motions under an identical scenario, both in intra-patient and inter-patient analyses, helps us to obtain a general pattern for the movement of different teeth using patient-specific teeth and bone geometries.

3 Setting up the Finite Element Model

This section describes different consecutive steps that are conducted to generate a patient-specific FEM of the human mandible. First, the geometry reconstruction takes place by segmenting the CBCT scan of the patient. Second, the surface mesh of the obtained geometries are re-meshed and a volumetric mesh is generated for each geometry. Next, the resulting volumetric meshes are imported into a finite element (FE) framework to set up the FE problem. The details of the biomechanical model, e.g., boundary conditions, contact definitions, and utilized material models are presented in this section. Finally, the model is numerically verified by using mesh convergence study and parameter sensitivity analysis.

Segmentation is performed using 3D Slicer [11] based on a semi-automatic watershed algorithm applied to the bone and teeth. Next, the wrongly segmented regions are modified to obtain the final segmentation result. Since the resolution of the orthodontic scans with a voxel size of $0.3 \times 0.3 \times 0.3$ mm^3 is not high enough for segmenting the thin PDL tissue (≈ 0.2 mm width) from the scans, the PDL layer is generated with a uniform width of 0.2 mm around each tooth root as shown in Fig. 1. We select three patients' scans of various crown height, root length, and teeth sizes, to ensure having enough geometrical variations. Each segmentation result is later verified by an orthodontic expert.

The segmented geometries are exported as surface meshes in STL files. These meshes are decimated and re-meshed using Meshmixer [2] to provide high-quality surface meshes. Uniform meshes are used for teeth and PDL geometries. Table 1 presents the edge length of the triangular meshes for each component. For bone geometry, an adaptive mesh is generated in which the edge length of the surface mesh triangles varies between 0.4 and 2 mm from the neighboring regions to the PDL and the bottom region of the mandible. Utilizing an adaptive mesh helps us to obtain a finer mesh in the regions of interest, and consequently, an accurate result in the FE analysis, yet reducing the total number of elements.

High quality volumetric meshes are generated for each surface mesh using Tet-Gen [33], by defining an upper limit for the radius-edge ratio of to-be-generated tetrahedra. This mesh quality constraint controls the ratio between the radius of the circumscribed sphere and the shortest edge of each tetrahedron, which prevents the production of low-quality (badly shaped) tetrahedra. Later, four different mesh quality measurements presented in [30], i.e., the volume-edge ratio [21, 25], the radius-edge ratio [3], and the radius ratio [4, 12] are chosen to verify the quality of the generated 4-noded tetrahedral meshes (TET4) (see the quality histograms

Table 1 Summary of the materials and mesh properties

	Material model	Material properties		Mesh Properties	
		Young's modulus (MPa)	Poisson's ratio (-)	Surface mesh edge length (mm)	Number of tetrahedra
Tooth	Rigid body	–	–	0.4	8,000
Bone	Isotropic elastic	1.5×10^3	0.3	Adaptive mesh (from 0.4 to 2)	3,255,000
		C_1 (MPa)	C_2 (MPa)		
PDL	Mooney-Rivlin [a]	0.011875	0	0.1	90,700

[a]$C_2 = 0$ reduces the Mooney-Rivlin material to uncoupled Neo-hookean. The values assigned for C_1 and C_2 correspond to the Young's modulus and Poisson's ratio of 0.0689 MPa and 0.45, respectively

in Fig. 1). Finally, the obtained meshes are used to set up and solve the FE problem. For reproducibility, we generate and solve the computational biomechanical models in FEBio software package [10] which is an open-source software for nonlinear FEA in biomechanics. A nonlinear quasi-static simulation is performed to analyze the teeth displacements in each FE model.

The different domains of the FEM, material properties, contact types, boundary conditions and the applied load are summarized in Fig. 2. To simplify the proposed model, the tooth domain is assumed as rigid-body with 6 degrees of freedom. The center of mass for each tooth is calculated automatically using FEBio based on a predefined density parameter [22]. Furthermore, since the deformation of the bone tissue is negligible under the orthodontic forces, no distinction is made for the cortical and trabecular bone [28, 34], and an isotropic elastic material model is used for the homogeneous bone geometry.

The importance of the PDL tissue in transferring loads from the tooth to the alveolar bone has been shown in the literature [5, 23, 26]. Accordingly, the PDL tissue is included in our model as a thin layer of finite elements [5, 14, 26, 29]. This allows for investigating the stress/strain field in the PDL, e.g., using data-driven models, that can later be used in the bone remodeling process [7]. Moreover, several studies have characterized the biomechanical behavior of the PDL tissue [9, 28, 32], some of which have suggested the Mooney-Rivlin Hyperelastic (MRH) model for the PDL [28, 32]. In this study, an MRH material model is used for the PDL domain based on the parameter values reported in Table 1.

The Tooth-PDL interface and PDL-Bone interface are fixed in both normal and tangent directions using a Neuman condition (see Fig. 2). In addition, all elements at the bottom surface of the bone (Γ_D) are fixed in all directions by applying a Dirichlet boundary condition.

To mimic the uncontrolled tipping scenario, a pressure load is applied perpendicular to the labial/buccal surface of the tooth crown, as shown in Fig. 2. The area under the load, which represents the area under the orthodontic bracket, is set to the center of the teeth crowns. To ensure that the same force magnitude is applied to the teeth, the area under the load is measured separately for each tooth. Next, the corresponding pressure value for the desired force magnitude is calculated and used as the *pressure load* in FEBio. An identical force magnitude is exerted to all teeth simultaneously in order to investigate the tooth movement variations of the mandibular teeth across the three patients.

The model is then verified by studying the mesh convergence and parameters sensitivities. The final resolution of the mesh is defined in the mesh convergence study process where the total number of elements, except for the rigid body teeth, is iteratively increased by a factor of 2 until the relative error is less than 4% of the maximum stress (see Fig. 3). The number of tetrahedra in the refined mesh is presented in Table 1. The parameter sensitivity study, summarized in Table 2, is done on the material parameters of different tissues and the parameters used for the tied contact in the PDL-bone interface.

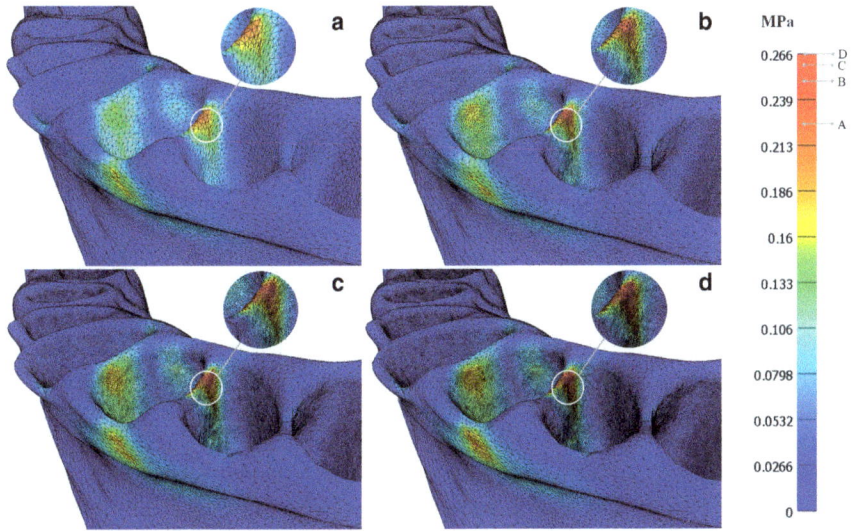

Fig. 3 Mesh convergence study showing the Von Mises stress in the bone geometry under the same boundary conditions. **a** to **d**: The model with N (coarse), $2N$, $4N$, and $8N$ elements. The stress fields are consistent in the finer models

Table 2 Summary of parameter sensitivity analysis conducted on the model

	PDL		Bone		PDL-Bone tied contact	
	Young's modulus	Poisson's ratio	Young's modulus	Poisson's ratio	Augmented lagrangian	penalty factor
Interval of parameter change	0.044–0.0938	0.45–0.49	1200–13700	0.2–0.4	0.2–0.1	0.25–1.75
Relative difference in Von Mises stress (%)	2.127	2.127	0.709	3.900	1.418	3.900

4 Experiments and Results

This section describes the experimental setup under the uncontrolled tipping scenario among the three patients. First, the obtained results are presented and discussed. Next, we propose two functions that describe the translation and rotation of different tooth types of the patients based on tooth IDs and selected clinical biomarkers, i.e., crown height and root volume.

In order to have a comprehensive inter-patient analysis, we select the patients with roughly the same number of teeth. We use the intraoral scan of each patient, captured by the 3Shape Trios scanner [1], to obtain the crown height of each tooth. Therefore, we ensure that the intraoral optical scans are available for all patients, and

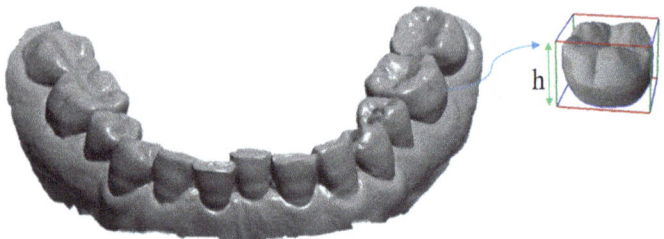

Fig. 4 An intraoral scan of a patient and the obtained crown height for a tooth

the CBCT scans have sufficient quality for performing the segmentation. Figure 4 shows an intraoral scan of a patient and illustrates how the crown height is obtained for a tooth.

In each scenario, an identical force magnitude is applied perpendicular to the surface of each tooth. The load magnitude (l) varies from 0.3 N to 1 N with 0.1 N intervals. The displacement of each tooth is measured as the translation of center of the mass ($\vec{t}$) and rotation of the rigid body teeth (with angle θ and axis $\vec{n}$). Besides, in each simulation, we record the tooth ID (k), load magnitude, and the relevant biomarkers. *Universal Numbering (UNN)* system is used for the tooth ID, where k changes between 17 and 32 from the left third molar to the right third molar, respectively. Figure 5 illustrates relation between the translation/rotation and applied load for different teeth of each patient. As can be seen, the translation magnitude of the mandibular incisors for an applied load of 0.4 N changes from 0.07 mm to 0.11 mm, 0.13 mm to 0.18 mm, and 0.24 mm to 0.51 mm, for patient 1 through 3, respectively. These values are similar to the results of the clinical study done by Jones et al. [16], where the obtained initial tooth movements ranged from 0.012 mm to 0.133 for maxillary incisors of ten patients under a constant load of 0.39 N over one-minute cycles.

The translation magnitude and rotation angle of the teeth can be described as the square root functions of the applied load, i.e., $t_{j,k} = \alpha_{t_{j,k}} \sqrt{l} + \beta_{t_{j,k}}$ and $\theta_{j,k} = \alpha_{\theta_{j,k}} \sqrt{l} + \beta_{\theta_{j,k}}$, where $\alpha_{t_{j,k}}$ and $\alpha_{\theta_{j,k}}$ are the translation/rotation function coefficients for the k-th tooth of the j-th patient, and $\beta_{t_{j,k}}$ and $\beta_{\theta_{j,k}}$ are the corresponding function intercepts which are nearly zero. This nonlinear relation between the displacement and load is in line with the experimental findings of the clinical study of [8] and numerical results of the biomechanical model of [6]. However, the function coefficients vary across different patients' teeth, i.e., the values increase when moving from the molars to central incisors.

It can be deduced from Fig. 5 that the estimated coefficients of the nonlinear functions vary across different patients for the same tooth types. This finding is due to the fact that the initial tooth movement can be influenced by different factors such as tooth anatomical variations and surrounding alveolar bone and PDL layer. Additionally, the root length of tooth and its surrounding alveolar bone can affect the initial tooth movement, center of rotation, and center of the resistance [31]. The same

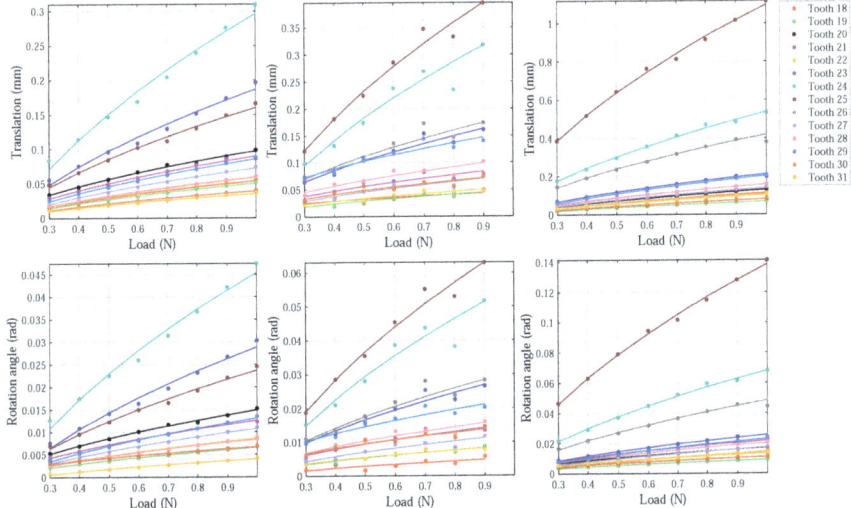

Fig. 5 The translation magnitudes and rotation angles versus the applied load. A nonlinear square root regression model is applied to fit the data from each patient's tooth. **Top-row**: Translation magnitude, **Bottom-row**: Rotation angle, **Left to Right**: Patient 1 through 3

behavior applies for the crown height. In other words, a specific tooth with a longer crown (or a shorter root) would experience more displacements than the same tooth with a shorter crown (or a longer root). However, the exact relationship between the crown/root size and tooth displacement is missing. Our hypothesis is that the intra- and inter-patient variations in crown and root size can influence the teeth movements of different patients. Therefore, we propose the ratio of crown height to root volume as the biomarker causing tooth movement variations together with the applied load.

To investigate the abovementioned assumption, first, we analyze the estimated coefficients of the fit functions ($\alpha_{t_{j,k}}$ and $\alpha_{\theta_{j,k}}$) for each tooth type. We observe that the teeth on the right side of the mandible show the same movement patterns as the corresponding teeth on the left side, where the UNN of the left side teeth can be calculated by subtracting the UNN of the corresponding teeth on the right side from 49. This provides us with more data points for the fitting purpose. The estimated coefficients of the nonlinear translation-load functions of the different patients are shown per tooth ID in Fig. 6. Note that the right teeth IDs are reflected in the same plot using the corresponding left teeth IDs.

Second, the crown heights of teeth are extracted from the intraoral scan of each patient. These measured values are then divided by the root volumes of the corresponding teeth, in which the root volumes are calculated using the associated bounding boxes of the PDL geometries. Figure 6 illustrates the estimated biomarker values for each patient's tooth. The obtained ratios needs to be considered as a patient's tooth biomarker in the tooth displacement models of translation and rotation. Therefore, we investigate the relationship between the coefficients of the displacement

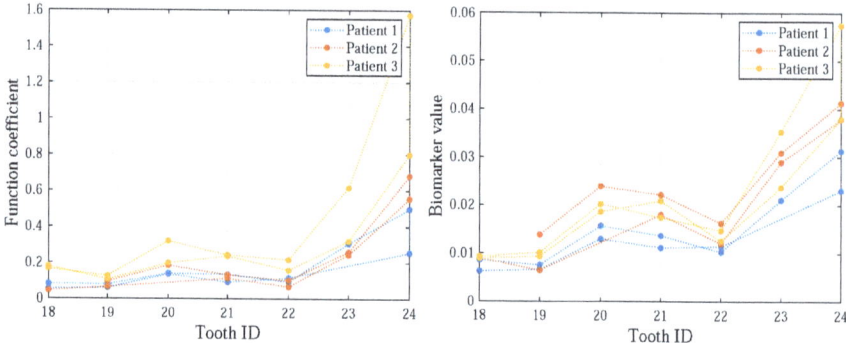

Fig. 6 Teeth movement variations of the three patients. **Left**: The coefficients of the functions fitted to the translation-load data. **Right**: The proposed biomarker values estimated for each patient's teeth. The right teeth IDs are reflected using the corresponding left teeth IDs, which results in two curves per patient

functions and the proposed biomarker values. Figure 7 shows the biomarker values versus coefficients of the teeth displacement functions (translation magnitude and rotation angle) for all patients' teeth. As it can be seen, the biomarker values and coefficients are in line with each other. For example, lower biomarker values and coefficients are associated with the molars while higher biomarker values and coefficients belongs to the incisors.

The relation between the biomarker values and coefficients of the teeth displacements can also be described by the square root functions, i.e., $b_t = \lambda_t \sqrt{\alpha_t} + \gamma_t$ and $b_\theta = \lambda_\theta \sqrt{\alpha_\theta} + \gamma_\theta$, where b_t and b_θ are the biomarker functions associated with the translation/rotation function coefficients α_t and α_θ, respectively. Hence, the tooth displacements (translation/rotation) will be seen as a nonlinear function of both load (l) and the proposed biomarker (b), wherein the displacement-load function coefficients are replaced with the biomarker values. In other words, to obtain a patient's tooth displacements $t_{j,k}$ and $\theta_{j,k}$ for an applied load, one only needs to obtain the function coefficients $\alpha_{t_{j,k}}$ and $\alpha_{\theta_{j,k}}$ based on the biomarker value of the specific tooth using the fits shown in Fig. 7.

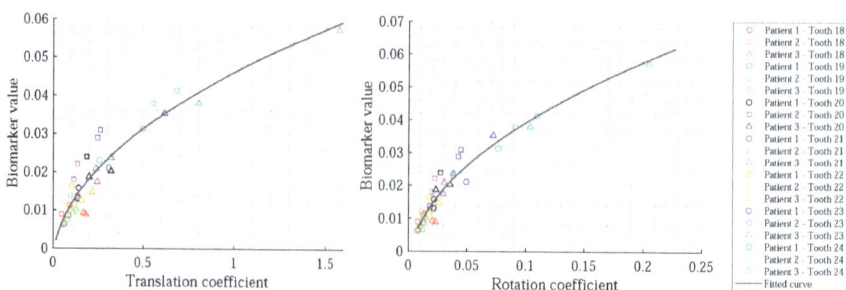

Fig. 7 The biomarker values versus coefficients of the displacement functions (translation and rotation). In each case, the behavior of the data is explained by a square root function

5 Summary and Conclusion

The main goal of this work was to introduce a computational analysis tool for investigating the influence of the teeth geometry of different patients on the resulting teeth movements. Three biomechanical models were generated for studying the tooth movement variations of three patients. Our study showed that a combination of two clinical biomarkers, i.e., crown height and root volume could affect the tooth displacement. Therefore, we proposed two nonlinear functions for predicting translation and rotation of different patients' teeth for any applied load magnitudes. Proposing such functions not only allows for generalizability of the model across different patients but also provides a way to avoid having multiple values for different teeth IDs. To the best of our knowledge, this is the first time a full dentition intra-patient and inter-patient tooth movement analyses have been considered. This study focused on modeling the movement of teeth under an uncontrolled tipping scenario applied to three patients. The work still can benefit from investigating different tooth movement types such as the crown tipping, root tipping, and pure translation applied to some more patients.

Acknowledgements This project has received funding from the European Union's Horizon 2020 research and innovation programme under the Marie Sklodowska-Curie grant agreement No. 764644. This paper only contains the author's views and the Research Executive Agency and the Commission are not responsible for any use that may be made of the information it contains.

References

1. 3Shape Trios Intraoral Scanner. https://www.3shape.com/en/scanners/trios.
2. Autodesk Meshmixer. http://www.meshmixer.com/.
3. Baker, T. J. (1989). Element quality in tetrahedral meshes. In *Proceedings of the 7th International Conference on Finite Element Methods in Flow Problems* (p. 1018).
4. Caendish, J. C., Field, D. A., & Frey, W. H. (1985). An apporach to automatic three-dimensional finite element mesh generation. *International Journal for Numerical Methods in Engineering, 21*(2), 329–347.
5. Cattaneo, P., Dalstra, M., & Melsen, B. (2005). The finite element method: A tool to study orthodontic tooth movement. *Journal of Dental Research, 84*(5), 428–433.
6. Cattaneo, P. M., Dalstra, M., & Melsen, B. (2008). Moment-to-force ratio, center of rotation, and force level: A finite element study predicting their interdependency for simulated orthodontic loading regimens. *American Journal of Orthodontics and Dentofacial Orthopedics, 133*(5), 681–689.
7. Chen, J., Li, W., Swain, M. V., Darendeliler, M. A., & Li, Q. (2014). A periodontal ligament driven remodeling algorithm for orthodontic tooth movement. *Journal of Biomechanics, 47*(7), 1689–1695.
8. Christiansen, R. L., & Burstone, C. J. (1969). Centers of rotation within the periodontal space. *American Journal of Orthodontics, 55*(4), 353–369.
9. Dorow, C., Krstin, N., & Sander, F. G. (2003). Determination of the mechanical properties of the periodontal ligament in a uniaxial tensional experiment. *Journal of Orofacial Orthopedics/Fortschritte der Kieferorthopädie, 64*(2), 100–107.
10. FEBio Software Suite. https://febio.org/.

11. Fedorov, A., Beichel, R., Kalpathy-Cramer, J., Finet, J., Fillion-Robin, J. C., Pujol, S., et al. (2012). 3D Slicer as an image computing platform for the quantitative imaging network. *Magnetic Resonance Imaging, 30*(9), 1323–1341.
12. Freitag, L. A., & Ollivier-Gooch, C. (1997). Tetrahedral mesh improvement using swapping and smoothing. *International Journal for Numerical Methods in Engineering, 40*(21), 3979–4002.
13. Hamanaka, R., Yamaoka, S., Anh, T. N., Tominaga, J. Y., Koga, Y., Yoshida, N. (2017) Numeric simulation model for long-term orthodontic tooth movement with contact boundary conditions using the finite element method. *American Journal of Orthodontics and Dentofacial Orthopedics* **152**(5), 601–612.
14. Hohmann, A., Kober, C., Young, P., Dorow, C., Geiger, M., Boryor, A., et al. (2011). Influence of different modeling strategies for the periodontal ligament on finite element simulation results. *American Journal of Orthodontics and Dentofacial Orthopedics, 139*(6), 775–783.
15. Huang, H. L., Tsai, M. T., Yang, S. G., Su, K. C., Shen, Y. W., & Hsu, J. T. (2020). Mandible integrity and material properties of the periodontal ligament during orthodontic tooth movement: A finite-element study. *Applied Sciences, 10*(8), 2980.
16. Jones, M., Hickman, J., Middleton, J., Knox, J., & Volp, C. (2001). A validated finite element method study of orthodontic tooth movement in the human subject. *Journal of Orthodontics, 28*(1), 29–38.
17. Kawamura, J., Park, J. H., Kojima, Y., Kook, Y. A., Kyung, H. M., & Chae, J. M. (2019). Biomechanical analysis for total mesialization of the mandibular dentition: A finite element study. *Orthodontics & Craniofacial Research, 22*(4), 329–336.
18. Li, Y., Jacox, L. A., Little, S. H., & Ko, C. C. (2018). Orthodontic tooth movement: The biology and clinical implications. *The Kaohsiung Journal of Medical Sciences, 34*(4), 207–214.
19. Liang, W., Rong, Q., Lin, J., & Xu, B. (2009). Torque control of the maxillary incisors in lingual and labial orthodontics: A 3-dimensional finite element analysis. *American Journal of Orthodontics and Dentofacial Orthopedics, 135*(3), 316–322.
20. Likitmongkolsakul, U., Smithmaitrie, P., Samruajbenjakun, B., & Aksornmuang, J. (2018). Development and validation of 3D finite element models for prediction of orthodontic tooth movement. *International Journal of Dentistry.*
21. Liu, A., & Joe, B. (1994). Relationship between tetrahedron shape measures. *BIT Numerical Mathematics, 34*(2), 268–287.
22. Maas, S., Weiss, J. (2007) FEBio user's manual. https://help.febio.org/FEBio/FEBio_um_2_8/index.html. [Online; Accessed 1 March 2019].
23. McCormack, S. W., Witzel, U., Watson, P. J., Fagan, M. J., & Gröning, F. (2014). The biomechanical function of periodontal ligament fibres in orthodontic tooth movement. *Plos One, 9*(7), e102387.
24. Melsen, B., Cattaneo, P. M., Dalstra, M., & Kraft, D. C. (2007) The importance of force levels in relation to tooth movement. In *Seminars in Orthodontics* (Vol. 13, pp. 220–233). Elsevier.
25. Misztal, M. K., Erleben, K., Bargteil, A., Fursund, J., Christensen, B. B., Bærentzen, J. A., & Bridson, R. (2013). Multiphase flow of immiscible fluids on unstructured moving meshes. *IEEE Transactions on Visualization and Computer Graphics, 20*(1), 4–16.
26. Ortún-Terrazas, J., Cegoñino, J., Santana-Penín, U., Santana-Mora, U., & del Palomar, A. P. (2018). Approach towards the porous fibrous structure of the periodontal ligament using micro-computerized tomography and finite element analysis. *Journal of the Mechanical Behavior of Biomedical Materials, 79,* 135–149.
27. Park, M., Na, Y., Park, M., & Ahn, J. (2017). Biomechanical analysis of distalization of mandibular molars by placing a mini-plate: a finite element study. *The Korean Journal of Orthodontics, 47*(5), 289–297.
28. Qian, L., Todo, M., Morita, Y., Matsushita, Y., & Koyano, K. (2009). Deformation analysis of the periodontium considering the viscoelasticity of the periodontal ligament. *Dental Materials, 25*(10), 1285–1292.
29. Savignano, R., Viecilli, R. F., Paoli, A., Razionale, A. V., & Barone, S. (2016). Nonlinear dependency of tooth movement on force system directions. *American Journal of Orthodontics and Dentofacial Orthopedics, 149*(6), 838–846.

30. Shewchuk, J. R. (2002). What is a good linear finite element? Interpolation, conditioning, anisotropy, and quality measures. *University of California at Berkeley, 73,* 137.
31. Tanne, K., Nagataki, T., Inoue, Y., Sakuda, M., & Burstone, C. J. (1991). Patterns of initial tooth displacements associated with various root lengths and alveolar bone heights. *American Journal of Orthodontics and Dentofacial Orthopedics, 100*(1), 66–71.
32. Uhlir, R., Mayo, V., Lin, P. H., Chen, S., Lee, Y. T., Hershey, G., et al. (2016). Biomechanical characterization of the periodontal ligament: Orthodontic tooth movement. *The Angle Orthodontist, 87*(2), 183–192.
33. WIAS-Software, TetGen. http://wias-berlin.de/software/index.jsp?id=TetGen&lang=1.
34. Ziegler, A., Keilig, L., Kawarizadeh, A., Jäger, A., & Bourauel, C. (2005). Numerical simulation of the biomechanical behaviour of multi-rooted teeth. *The European Journal of Orthodontics, 27*(4), 333–339.

Computational Biomechanics Model for Analysis of Cervical Spinal Cord Deformations Under Whiplash-Type Loading

Daniel Tan, Stuart I. Hodgetts, Sarah Dunlop, Karol Miller, Koshiro Ono, and Adam Wittek

Abstract Computational biomechanics analysis of neurologic injuries has so far been focused on the brain. This has resulted in the development of industrially applied computational biomechanics models of the brain and various criteria (such as cumulative strain damage measure, CSDM and maximum principal strain, MPS) to determine the likelihood of brain injuries. In contrast, spinal cord injury (SCI) has attracted relatively little attention from the computational biomechanics research community. This study, addresses this gap by applying computational biomechanics analysis to predict deformations of the cervical spinal cord and the risk of SCI. To do this, a three-dimensional finite element model of the cervical spinal cord was built using data from the Visible Human Project and incorporated into a well-established finite element model of the human body for injury analysis, created by the Global Human Body Models Consortium (GHMBC). The model was applied to simulate well-documented low-speed whiplash-type experiments previously conducted on volunteers by the Japan Automobile Research Institute and University of Tsukuba. The results suggest that direct application of CSDM and MPS criteria and the related injury risk curves leads to overestimation of SCI risk.

D. Tan (✉) · K. Miller · A. Wittek
Intelligent Systems for Medicine Laboratory, The University of Western Australia, Crawley, WA 6009, Australia
e-mail: daniel.tan@research.uwa.edu.au

S. I. Hodgetts
School of Human Sciences, The University of Western Australia, Crawley, WA 6009, Australia

Perron Institute for Neurological and Translational Science, Nedlands, WA 6009, Australia

S. Dunlop
School of Biological Sciences, The University of Western Australia, Crawley, WA 6009, Australia

K. Miller
Harvard Medical School, Boston, MA, USA

K. Ono
National Institute of Advanced Industrial Science and Technology (AIST) (Retired, Japan Automobile Research Institute: JARI), Tsukuba, Ibaraki, Japan

© The Author(s), under exclusive license to Springer Nature Switzerland AG 2021
K. Miller et al. (eds.), *Computational Biomechanics for Medicine*,
https://doi.org/10.1007/978-3-030-70123-9_4

45

Keywords Spinal cord injury · Injury criteria · Explicit dynamics finite element analysis · Computational biomechanics model

1 Introduction

Spinal cord injury (SCI) results from damage to the spinal cord, a fundamental part of the central nervous system connecting the brain to other parts of the human body. As a result, SCI can lead to losses in motor and sensory function throughout the body. The occurrence of SCI is not insignificant either—in 2016 there were 0.93 million new cases of SCI world-wide, adding to the existing global population of 27 million [1].

The impact of SCI correlates to the vertebral level of the injury—in humans ranging from cervical (8 levels) to thoracic (12 levels) to lumbar (5 levels) and finally sacral (5 levels). An injury to the cervical region of the spinal cord generally results in tetraplegia, which is the partial or total reduction or loss of motor and/or sensory function in the arms as well as in the trunk, legs and pelvic organs [2].

As SCI causes significant impact to the individual, but are challenging to analyse experimentally, a number of studies have used methods of computational biomechanics. Finite-element models for the cervical spinal cord have previously been created to simulate hyperextension injury [3] and burst fracture injury [4]. In addition to this, several finite element models incorporating cervical vertebrae together with the spinal cord have been reported [5, 6]. However, these models only focus on selected (isolated) cervical vertebrae and sections of the spinal cord.

The objective of this study is to predict the deformations of the cervical spinal cord and analyse the SCI risk for a given stimulus/acceleration impact load. To achieve this objective, we created a new three-dimensional (3D) finite element model of the entire cervical spinal cord (C1-C8) and incorporated it into the previously validated cervical spine model section of the full human body model created by the Global Human Body Models Consortium (GHBMC) [7]. To evaluate model performance, we attempted to simulate experiments conducted previously on volunteers subjected to whiplash-type loading at low-speed [8]. Because there are no existing specific injury criteria or quantitative measures to correlate spinal cord deformations with SCI risk, we used criteria that are well-established for brain injury namely, cumulative strain damage measure (CSDM) [9, 10] and maximum principal strain (MPS) [11, 12].

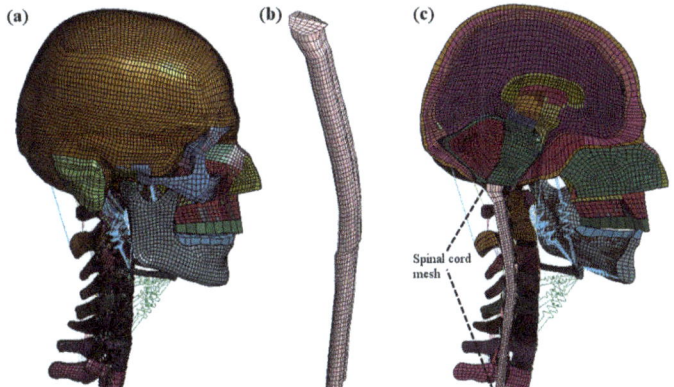

Fig. 1 **a** GHBMC cervical spine model; **b** Finite-element model of spinal cord created in study; **c** Midsagittal view of study model (with created spinal cord highlighted)

2 Methods

2.1 Cervical Spine and Spinal Cord Models

2.1.1 Cervical Spine Model

Cervical spine, head and brain models were extracted from version 4.3 of the full human body model created by the Global Human Body Models Consortium (GHBMC) [7]. These were implemented with LS-DYNA explicit dynamics finite element code [13] and referred to hereafter as the GHBMC cervical spine model (Fig. 1a). The extraction was done using HyperMesh version 12.0 [14] finite element mesh generator and LS-DYNA finite element code pre-processing and post-processing software LS-PrePost version 4.0 [15]. Reducing the model to the skull, brain, cervical vertebrae and associated structures, including cartilage, ligaments and intervertebral discs, appreciably shortened the computation time allowing us to conduct overnight simulations on an off-the-shelf personal computer.

2.1.2 Spinal Cord Model

To create the 3D geometry for the spinal cord part of the model, a series of high-resolution axial photographs (cryosections) from the U.S. National Library of Medicine Visible Human Project [16] were used as the data source. In the Visible Human Project, the axial photographs were taken at 1 mm intervals from a male cadaver with in-plane resolution of 2048×1216 pixels, which is sufficient to identify spinal cord grey and white matter. Because the quality (i.e. contrast) of the cryosection images was insufficient for automated segmentation, we undertook manual segmentation to obtain the spinal cord geometry using the 3D Slicer image computing software platform (www.slicer.org, [17]). This geometry data (discretised surfaces in stereolitography STL format) was then exported into Altair HyperMesh finite element mesh generator, where the finite element mesh was built (Fig. 1b). The created three-dimensional (3D) spinal cord mesh consists of 4740 eight-noded hexahedral elements (each with a single integration point) and 6160 nodes. To prevent hourglassing (zero-energy modes), we used the standard LS-DYNA viscous hourglass control [18], with an hourglass coefficient of 0.15.

2.1.3 Cervical Spine Model with Spinal Cord

The 3D spinal cord mesh from Fig. 1b was inserted into the GHBMC cervical spine model (Fig. 1a) to form the 3D model used in the study as shown in Fig. 1c.

2.1.4 Material Models and Properties

The material properties for this study were calibrated as described below to align with the responses measured in past experiments on human brain tissue specimens [19] subjected to tension and compression strain rates of 5/s consistent with low-severity whiplash-type impacts (Fig. 2).

As only limited experimental data about the biomechanical responses of spinal cord tissue under transient loads have been reported [29, 30], and both the brain and spinal cord consist of the same types of tissues (grey and white mater), for the spinal cord we used the same material properties as for the brain tissue.

Calibration of the material properties was undertaken through modelling of the experiments by Jin et al. [19]. Tissue samples were discretised using a mesh of 28 × 28 × 10 eight-noded hexahedral elements (each with one integration point) and standard LS-DYNA viscous hourglass control [18] (see Fig. 3). In total the mesh consisted of 7840 elements and 9251 nodes. As Jin et al. [19] used adhesive in the

Fig. 2 Brain tissue experimental schematic from Jin et al. [19]. The brain tissue sample (cuboid) dimensions were 14 mm × 14 mm × 5 mm

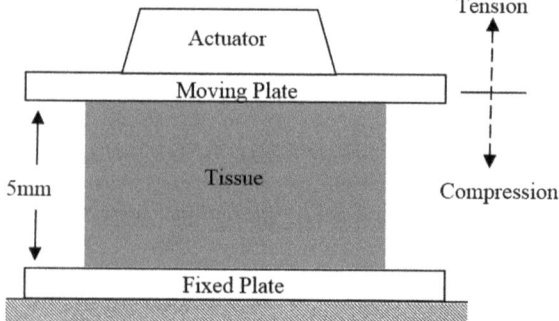

Fig. 3 Finite element model representing brain tissue used in experiments by Jin et al. [19]

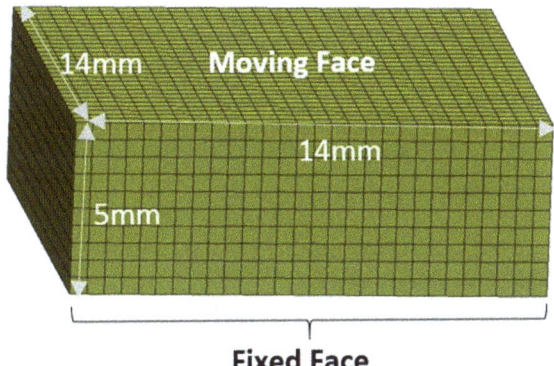

tension tests to attach the sample to the fixed and moving plates, all the nodes on the bottom model face in contact with the fixed plate were rigidly constrained, and the nodes on opposite (top) face in contact with the moving plate were allowed to move only in the vertical direction. The same constraints were also applied to the model of the compression test. Loading was imposed by prescribing a velocity of 25 mm/s (strain rate of 5/s) in the vertical direction to all nodes located on the top face of the sample.

Following the results of the experiments on brain tissue samples by Miller and Chinzei [20], the Ogden material model was used for both brain and spinal cord tissues:

$$W = \frac{2}{\alpha^2} \int_0^t \left[\mu(t-\tau)\frac{d}{d\tau}(\lambda_1^\alpha + \lambda_2^\alpha + \lambda_3^\alpha - 3) \right] d\tau + K(J - 1 - \ln J) \quad (1)$$

$$\mu = \mu_0 \left[1 - \sum_{k=1}^n g_k \left(1 - e^{-\frac{t}{\tau_k}} \right) \right] \quad (2)$$

where W is the strain energy function, λ_i (for $i = 1, 2, 3$) are the principal stretches, μ_0 is the instantaneous shear modulus, τ_k are the characteristic times, g_k are the relaxation coefficients, α is a material coefficient which can take any real value without restrictions [21], K is the bulk modulus and J is the relative change in volume.

The results of calibration of the Ogden material model parameters (as defined in Eqs. (1) and (2)), where $\beta = 1/\tau$, are shown in Table 1.

The dura mater surrounding the spinal cord was built from the outer surface/layer of spinal cord mesh and, following Wittek et al. [22], was assumed to have neo-Hookean material properties. As no dedicated functionality/feature for modelling of neo-Hookean material responses is available in LS-DYNA [13], we used Mooney-Rivlin material model with the parameters defined such that the behaviour is equivalent to the neo-Hookean material (Table 2):

Table 1 Parameters of the Ogden material model for the brain and spinal cord tissues obtained from calibration against the experiments by Jin et al. [19]. These properties were used in this study

Mass density (kg/m^3)	Poisson's ratio	Shear modulus, μ_0 (kPa)	Alpha, α	Shear relaxation modulus, g_i (kPa)	Decay constant, β_i (ms^{-1})
1.06×10^3	0.45	1.50	-4.7	7.50	0.0125

Table 2 Parameters for dura mater (Mooney-Rivlin material model, see Eq. (3)) used in this study

Mass density (kg/m^3)	Poisson's ratio	Parameter "A" (kPa)	Parameter "B" (kPa)
1.06×10^3	0.45	3.75	0

$$W = A(I - 3) + B(II - 3) + C\left(III^{-2} - 1\right) + D(III - 1)^2 \qquad (3)$$

where

$$C = 0.5A + B; \ D = \frac{A(5v - 2) + B(11v - 5)}{2(1 - 2v)}$$

v is Poisson's ratio, the shear modulus of linear elasticity is given by $2(A + B)$ and I, II, III are invariants of the right Cauchy-Green Tensor C.

2.1.5 Boundary Conditions for the Spinal Cord

In the model, the spinal cord was attached to the dura mater by implementing a tied contact in LS-DYNA finite element code [13] between the spinal cord and dura. Upon visual inspection of anatomical samples available from the Anatomy Department of The University of Western Australia, it was observed that the spinal roots extend outward from gaps in the dura to the physical gaps found between the vertebrae (human ethics approval constraints governing specimen use dictated only a schematic drawing (Fig. 4a) rather than a photograph can be shown). This was modelled by forcing nodes of the dura mater visible in the space between the two adjacent vertebrae, to follow the motion of the corresponding vertebrae (Fig. 4b) and implemented in the model through use of the LS-DYNA constrained node set function. For example in Fig. 4b, the C1 constrained node set (indicated as yellow points on the dura mesh) is constrained to the motion of the C1 vertebra.

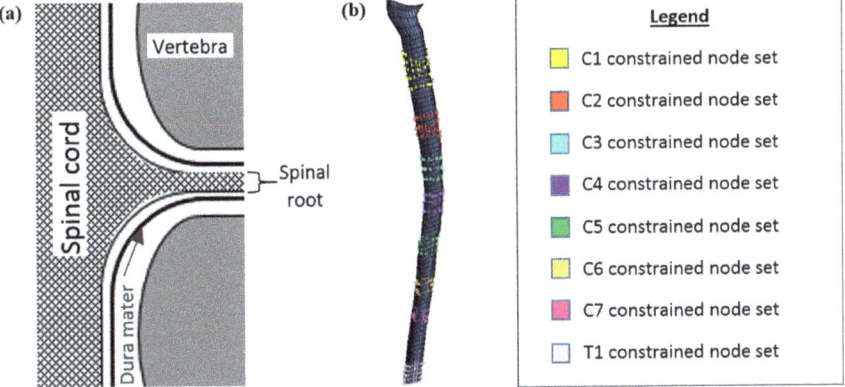

Fig. 4 **a** Schematic drawing of anatomical specimen showing spinal roots; **b** Boundary conditions for the spinal cord: constrained node sets on dura model accounting for the constraining role/effects of the spinal roots

2.2 Demonstration of the Model Application to Evaluation of Spinal Cord Injury Risk: Modelling of Whiplash-Type Experiments

The 3D cervical spine with spinal cord model created in this study (Fig. 1c) was used to simulate low-speed whiplash-like experiments previously undertaken by Ono et al. [8] (Fig. 5). External load was imposed by prescribing motion of the skull and vertebrae (all cervical vertebrae C1-C7) using the kinematic data of the cervical spine and head kinematics recorded using X-ray cineradiography [8]. These data sets, as plotted in Fig. 6, prescribed the time history of translation in the direction of X and

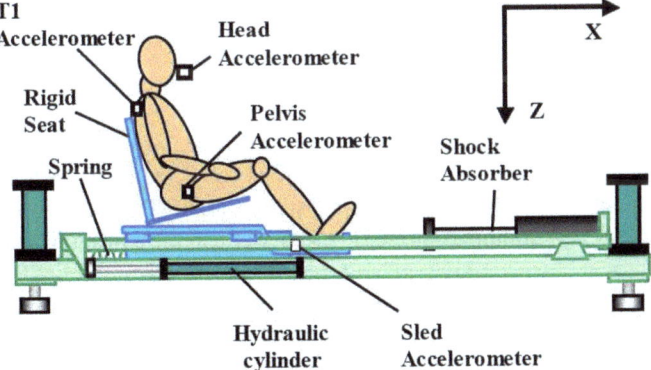

Fig. 5 Experimental setup of low-speed whiplash-like experiments, with Y-axis direction coming out of page. Adapted from Ono et al. [8]

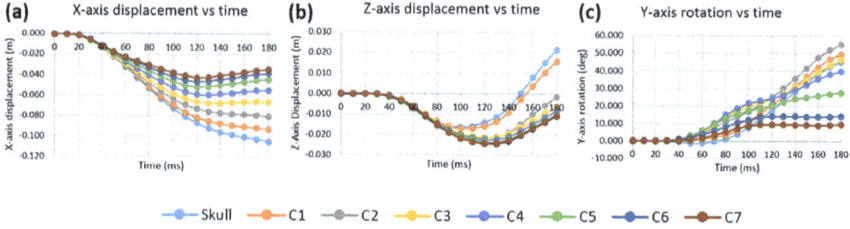

Fig. 6 Kinematic data of cervical vertebrae **a** X-axis translation; **b** Z-axis translation; and **c** Y-axis rotation from the experiments by Ono et al. [8]

Z-axes and rotation about the Y-axis, with the X, Y and Z axes defined as in Fig. 5. Defining the external load by prescribing motion of the boundary has the important advantage of ensuring that the predicted tissue deformations are only weakly sensitive to assumptions and inaccuracies regarding the tissue material models and material properties [23, 24].

As no widely accepted specialised criteria for the quantification of SCI risk has been proposed so far, we used two well-established brain injury criteria instead: maximum principal strain (MPS) [11, 12] and cumulative strain damage measure (CSDM) [9, 10]. To reduce the effects of modelling artefacts, we use 99th percentile of the maximum principal strain rather than raw strain when reporting MPS.

CSDM describes the volumetric ratio (percentage) of tissue subjected to strain above a certain level to the total volume of tissue of interest. Unlike the original CSDM formulation by Bandak and Eppinger [9], we determined this ratio not for the brain but for the spinal cord tissue. As for the strain measure, we used the maximum principal Almansi strain. Following Takhounts et al. [25], the strain threshold of 0.25 was used when computing CSDM.

3   Results

3.1   Calibration of Material Properties

The brain tissue sample model (from Fig. 3) was subjected to 50% elongation and 15% compression. However, hourglassing started after 9% compression and was clearly visible at 15% compression (see Fig. 7b), whereas the tension simulation ran to 50% elongation without any signs of hourglassing and other instabilities (see Fig. 7a). Quantitative comparison of the predicted nominal tensile and compressive stress values, and those experimentally determined by Jin et al. [19] (Figs. 8 and 9, Tables 3 and 4), confirm calibration validity and accuracy. Due to element distortion, we were unable to model compression exceeding 15% of the initial sample height. This highlights the limitations of finite element analysis in modelling continua undergoing large deformations and strains reported in our previous studies [26]. As the spinal

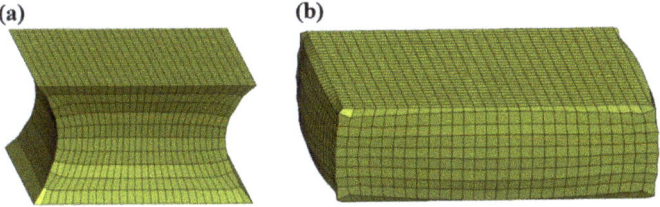

Fig. 7 Calibration of tissue material properties: modelling of the experiments by Jin et al. [19] **a** Brain tissue sample subjected to 50% elongation; **b** Brain tissue sample subjected to 15% compression

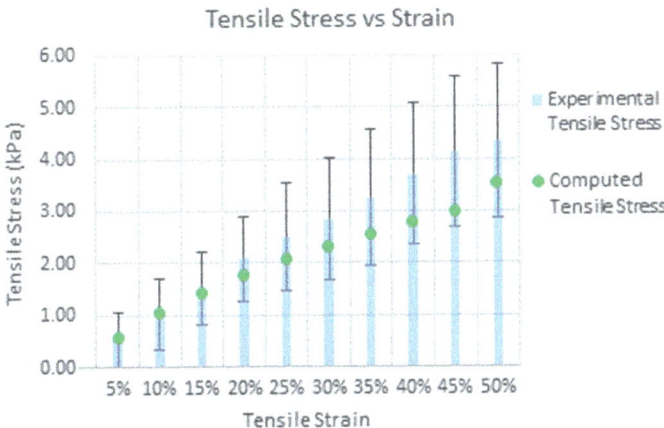

Fig. 8 Calibration of tissue material properties: comparison of the predicted and experimentally measured by Jin et al. [19] nominal tensile stress for a given nominal tensile strain

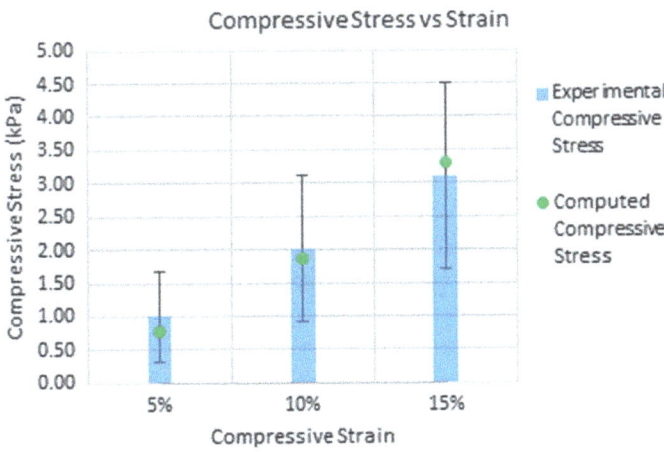

Fig. 9 Calibration of tissue material properties: comparison of the predicted and experimentally measured by Jin et al. [19] nominal compressive stress for a given nominal compressive strain

Table 3 Calibration of tissue material properties: Comparison of the predicted and experimentally measured by Jin et al. [19] nominal tensile stress for a given nominal strain. Std Dev. = standard deviation

Tensile strain (%)	Computed tensile stress (kPa)	Average experimental tensile stress (kPa)	Experimental tensile stress—Std Dev. (kPa)
5	0.584	0.533	0.522
10	1.020	1.029	0.667
15	1.404	1.526	0.700
20	1.733	2.083	0.828
25	1.998	2.501	1.026
30	2.249	2.844	1.183
35	2.482	3.255	1.296
40	2.702	3.711	1.348
45	2.914	4.137	1.443
50	3.436	4.335	1.478

Table 4 Calibration of tissue material properties: Comparison of the predicted and experimentally measured by Jin et al. [19] nominal compressive stress for a given nominal strain. Std Dev. = standard deviation

Compressive strain (%)	Computed compressive stress (kPa)	Average experimental compressive stress (kPa)	Experimental compressive stress Std Dev. (kPa)
5	0.784	1.000	0.688
10	1.869	2.012	1.096
15	3.329	3.105	1.403

cord in the whiplash-type experiments modelled and analysed here, was mainly undergoing tension rather than compression, calibration for compression of up to 15% should be regarded as sufficient.

3.2   Demonstration of the Model Application to Evaluation of Spinal Cord Injury Risk: Modelling of Whiplash-Type Experiments

We successfully modelled the entire 180 ms duration of the experiments by Ono et al. [8] (see Fig. 10). The 99th percentile of the MPS was computed as 0.39. The CSDM for the strain threshold of 0.25, CSDM (0.25) hereafter referred as CSDM, was computed to be 0.44. The predicted volume of the spinal cord tissue experiencing strain of 0.25 and above was 5529 mm^3 (out of the total spinal cord volume of 12618 mm^3).

Fig. 10 Mid-sagittal view of the 3D cervical spine with spinal cord model created and used in this study at 180 ms of simulation of the experiments by Ono et al. [8]

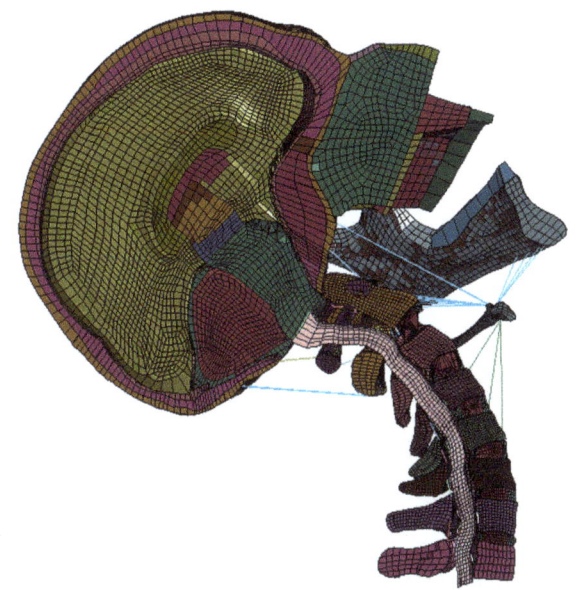

Whilst our model generates interesting results regarding the deformations of the cervical spinal cord under whiplash-type loading, the available experimental data do not allow direct quantitative validation of the results obtained. However, it may be possible to validate the results indirectly through prediction of the SCI risk using CSDM and MPS and comparing the predicted injury risk with the fact that none of the volunteers participating in the experiments by Ono et al. [8] suffered from pathological SCI and none even experienced neck pain.

As mentioned previously, our simulation of Ono et al. experiments [8] predicted CDSM of 0.44. For the well-established brain injury risk curves by Takhounts et al. [25], this corresponds to 85% risk of Abbreviated Injury Scale [27] AIS2 injury (moderate injury [28]), 55% risk of AIS3 injury (serious injury [28]), and 44% risk of an AIS4 injury (severe injury [28]) as shown in Fig. 11. These point to appreciable risk of moderate to severe spinal cord injury, and are consistent with the injury risk evaluation using MPS. Our simulation of the experiments by Ono et al. [8] also predicted 99th percentile MPS of 0.39. For the injury risk curves by Takhounts et al. [25], this corresponds to 40% risk of AIS2 injury, 11% risk of an AIS3 injury, and 7% risk of an AIS4 injury (Fig. 12). Therefore, the results obtain using CSDM and MPS are inconsistent with the fact no (even low severity) SCI to volunteers were observed in experiments by Ono et al. [8].

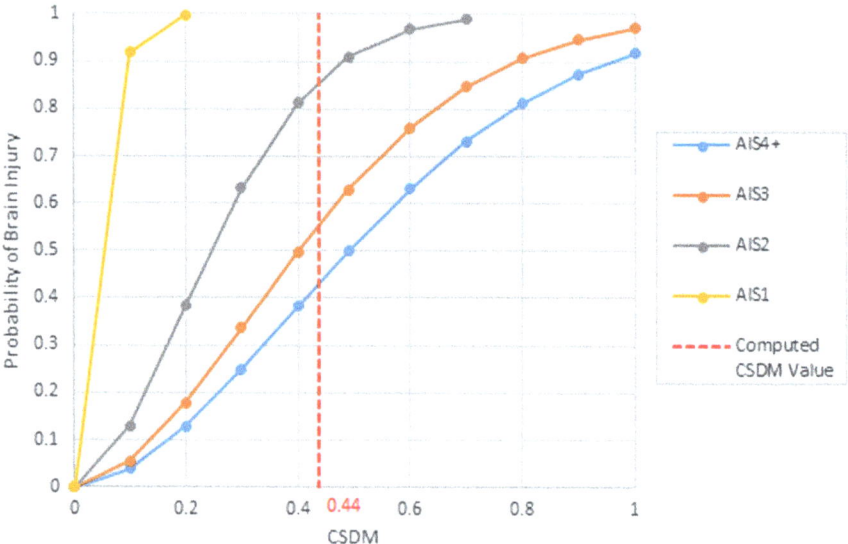

Fig. 11 Spinal cord Cumulative Strain Damage Measure (CSDM) predicted in this study when modelling experiments by Ono et al. [8] versus the CSDM brain injury risk curves by Takhounts et al. [25]

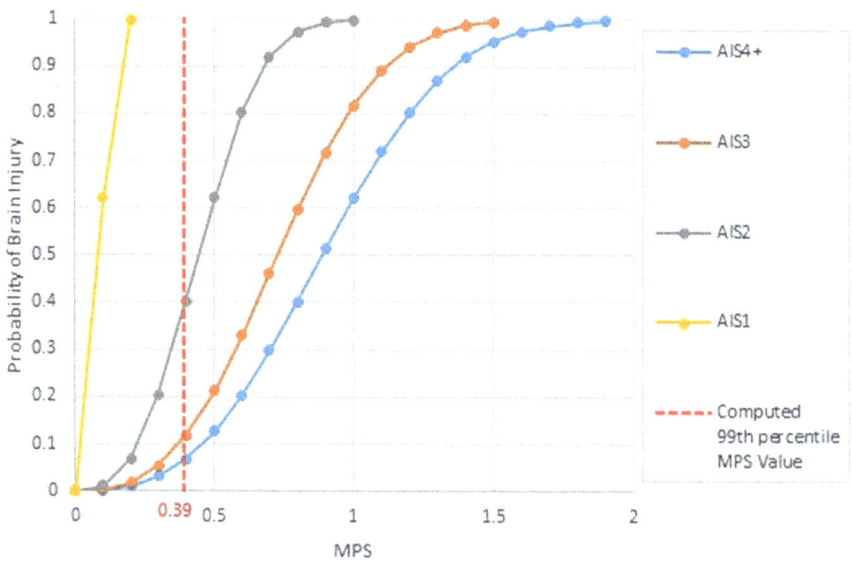

Fig. 12 Spinal cord 99th percentile Maximum Principal Strain (MPS) predicted in this study when modelling experiments by Ono et al. [8] versus the MPS brain injury risk curves by Takhounts et al. [25]

4 Discussion

The simulations of the brain tissue calibration tests suggest that the shear modulus of 1.50 kPa (and other parameters reported in Table 1) are suitable for setting as the material properties of brain tissue for strain rates of an order of 5/s. However, the instabilities encountered in the simulation for the brain tissue samples subjected to compression for strains greater than 15% highlight that the finite-element method may not be suitable for modelling of large compression, and another method of computational mechanics, such as meshless methods [26], should be used in these scenarios.

While we succeeded in calculating the MPS and CSDM for the cervical section of the 3D spinal cord model, the risk of cervical SCI predicted by using the calculated MPS and CSDM from the injury risk curves developed by Takhounts et al. [25] was inconsistent with the experimental observations. This suggests the need for developing specialised injury criteria and establishing new injury risk curves for the spinal cord.

Acknowledgements The academic license of the Global Human Body Models Consortium model was used for this study. Daniel Tan gratefully acknowledges the support of the Australian Postgraduate Awards scholarship to support this work.

References

1. James, S. L., Theadom, A., Ellenbogen, R. G., Bannick, M. S., Montjoy-Venning, W., Lucchesi, L. R., et al. (2019). Global, regional, and national burden of traumatic brain injury and spinal cord injury, 1990–2016: A systematic analysis for the global burden of disease study 2016. *Lancet Neurology, 18*(1), 56–87.
2. Australian Institute of Health and Welfare. (2020). *Spinal cord injury, Australia, 2016–17.* Canberra: AIHW.
3. Li, X. F., & Dai, L. Y. (2009). Three-dimensional finite element model of the cervical spinal cord: Preliminary results of injury mechanism analysis. *Spine, 34*(11), 1140–1147.
4. Czyz, M., Scigala, K., Jarmundowicz, W., & Bedzinski, R. (2008). The biomechanical analysis of the traumatic cervical spinal cord injury using finite element approach. *Acta of Bioengineering and Biomechanics, 10*(1), 43–54.
5. Greaves, C. Y., Gadala, M. S., & Oxland, T. R. (2008). A three-dimensional finite element model of the cervical spine with spinal cord: An investigation of three injury mechanisms. *Annals of Biomedical Engineering, 36*(3), 396–405.
6. Scifert, J., Totoribe, K., Goel, V., & Huntzinger, J. (2002). Spinal cord mechanics during flexion and extension of the cervical spine: a finite element study. *Pain Physician, 5*(4), 394–400.
7. Park, G., Kim, T., Crandall, J. R., Arregui-Dalmases, C., & Luzón-Narro, B. J. (2013). Comparison of kinematics of GHBMC to PMHS on the side impact condition. *Proc IRCOBI, 2013,* 368–379.
8. Ono, K., Ejima, S., Suzuki, Y., Kaneoka, K., Fukushima, M., & Ujihashi, S. (2006). Prediction of neck injury risk based on the analysis of localized cervical vertebral motion of human volunteers during low-speed rear impacts. *Proc IRCOBI, 2006,* 103–113.

9. Bandak, F. A., Eppinger, R. H. (1995) A three-dimensional finite element analysis of the human brain under combined rotational and translational acceleration. In: *Proceedings of the 38th Stapp Car Crash Conference* (pp. 145–163).
10. Takhounts, E. G., Eppinger, R. H., Campbell, J. Q., Tannous, R. E., Power, E. D., & Shook, L. S. (2003). On the development of the SIMon finite element head model (No. 2003-22-0007). *Stapp Car Crash J, 47,* 107–133.
11. Cloots, R. J. H., Gervaise, H. M. T., Van Dommelen, J. A. W., & Geers, M. G. D. (2008). Biomechanics of traumatic brain injury: Influences of the morphologic heterogeneities of the cerebral cortex. *Annals of Biomedical Engineering, 36*(7), 1203.
12. McAllister, T. W., Ford, J. C., Ji, S., Beckwith, J. G., Flashman, L. A., Paulsen, K., et al. (2012). Maximum principal strain and strain rate associated with concussion diagnosis correlates with changes in corpus callosum white matter indices. *Annals of Biomedical Engineering, 40*(1), 127–140.
13. Livermore Software Technology Corporation (LSTC). (2020). LS-DYNA version 971 R6.0. http://lstc.com/products/ls-dyna.
14. Altair Engineering. (2020). Altair HyperMesh version 12.0. https://www.altair.com/hypermesh/.
15. Livermore Software Technology Corporation (LSTC). (2012). LS-PrePost version 4.0 https://www.lstc.com/products/ls-prepost.
16. Ackerman, M. J. (1998). The visible human project. *Proceedings of the IEEE Institution of Electrical and Electronics Engineers, 86*(3), 504–511.
17. Fedorov, A., Beichel, R., Kalpathy-Cramer, J., Finet, J., Fillion-Robin, J. C., Pujol, S., et al. (2012). 3D Slicer as an image computing platform for the Quantitative Imaging Network. *Magnetic Resonance Imaging, 30*(9), 1323–1341.
18. Livermore Software Technology Corporation (LSTC). (2020). LS-DYNA: Keywords user manual–Volume I. https://www.dynasupport.com/manuals/ls-dyna-manuals/ls-dyna_manual_volume_i_r12.pdf.
19. Jin, X., Zhu, F., Mao, H., Shen, M., & Yang, K. H. (2013). A comprehensive experimental study on material properties of human brain tissue. *Journal of Biomechanics, 46*(16), 2795–2801.
20. Miller, K., & Chinzei, K. (2002). Mechanical properties of brain tissue in tension. *Journal of Biomechanics, 35*(4), 483–490.
21. Wittek, A., Miller, K., Kikinis, R., & Warfield, S. K. (2007). Patient-specific model of brain deformation: Application to medical image registration. *Journal of Biomechanics, 40*(4), 919–929.
22. Wittek, A., Dutta-Roy, T., Taylor, Z., Horton, A., Washio, T., Chinzei, K., et al. (2008). Subject-specific non-linear biomechanical model of needle insertion into brain. *Computer Methods in Biomechanics and Biomedical Engineering: Imaging & Visualization, 11*(2), 135–146.
23. Wittek, A., Hawkins, T., & Miller, K. (2009). On the unimportance of constitutive models in computing brain deformation for image-guided surgery. *Biomechanics and Modeling in Mechanobiology, 8*(1), 77–84.
24. Miller, K., & Lu, J. (2013). On the prospect of patient-specific biomechanics without patient-specific properties of tissues. *Journal of the Mechanical Behavior of Biomedical Materials, 27,* 154–166.
25. Takhounts, E. G., Craig, M. J., Moorhouse, K., McFadden, J., & Hasija, V. (2013). Development of brain injury criteria (Br IC) (No. 2013–22-0010). *Stapp Car Crash Journal, 57,* 243–266.
26. Joldes, G., Bourantas, G., Zwick, B., Chowdhury, H., Wittek, A., Agrawal, S., et al. (2019). Suite of meshless algorithms for accurate computation of soft tissue deformation for surgical simulation. *Medical Image Analysis, 56,* 152–171.
27. Abbreviated Injury Scale. (2005). *(2008 Updated) Association for the Advancement of Automotive Medicine (AAAM.* Illinois: Des Plaines.
28. Radja, G. A. (2016). National automotive sampling system–crashworthiness data system (NASS-CDS), 2015 analytical user's manual (No. DOT HS 812 321).

29. Bilston, L. E., & Thibault, L. E. (1995). The mechanical properties of human cervical spinal cord in vitro. *Annals of Biomedical Engineering, 24*(1), 67–74.
30. Shi, R., & Pryor, R. D. (2002). Pathological changes of isolated spinal cord axons in response to mechanical stretch. *Neuroscience, 110*(4), 765–777.

Biomechanical Tissue Characterisation, Determining Organ Geometry, and Organ Deformation Measurements

An Unsupervised Learning Based Deformable Registration Network for 4D-CT Images

Dongming Wei, Wenlong Yang, Pascal Paysan, and Hefeng Liu

Abstract Four-dimensional computed tomography (4D-CT) has been used in radiation therapy, which allows for tumor and organ motion tracking through the breathing cycle. Based on the motion trajectory analysis of tumor and normal tissues, an adaptive treatment planning may be improved in terms of the accuracy of tumor delineation, or gating radiation. Image registration can be used to compensate the motion and supply the transformation between different scans in 4D-CT to help further analysis. The motion in the 4D-CT is mainly caused by the respiration, which requires deformable image registration. As the accurate and fast function facilitated by GPU, deep learning based deformable image registration methods are widely used for 4D-CT. In this paper, we apply a deep learning based deformable registration network (Reg-Net) to estimate the deformation field for the given scan pair of 4D-CT. The proposed network was trained in an unsupervised manner without the need of any expert annotation. For evaluation, 10 subjects are used for training, 5 subjects for testing. We use the Dice similarity coefficient (DSC) and intersection over union (IoU) over regions of interest (*i.e.,* gross target volume) to evaluate the registration performance. The experimental results demonstrate that the proposed deformable Reg-Net can potentially improve the organ tracking performance.

Keywords Radiation therapy · 4D CT · Deformable registration · Unsupervised learning

D. Wei · W. Yang (✉) · H. Liu
Varian Medical Systems, Inc., Shangai, China
e-mail: 290125097@99.com

D. Wei
e-mail: dongming.wei@sjtu.edu.cn

P. Paysan
Varian Medical Systems, Inc., Steinhausen, Switzerland

© The Author(s), under exclusive license to Springer Nature Switzerland AG 2021
K. Miller et al. (eds.), *Computational Biomechanics for Medicine*,
https://doi.org/10.1007/978-3-030-70123-9_5

63

1 Introduction

Respiration-induced abdominal and thoracic tissue motion [1] renders the difficulty in accurate treatment planning and radiation therapy (RT) for abdominal and thoracic cancer. Four dimensional computed tomography (4D-CT) has been used as clinical standard in RT for treatment planning, so as to reduce dose to healthy organs and increase dose to the tumor target [2, 3]. As the fast imaging speed in 4D-CT, it is commonly used to evaluate and manage patient respiratory motion, especially for the lung, abdominal, and thoracic cancers [4]. An exemplar of 4D-CT can be seen in Fig. 1, where ten time-point phases in a respiration circle and one free breathing scan are involved. The lung motion can be clearly seen among ten phases.

Deformable image registration (DIR), which builds the anatomical correspondences among different images, is a key technology for many applications in RT. A recent development [5] applies a fast and accurate intra-patient DIR method to transfer the therapy plan. There are also several works [6, 7] to build motion models that represent the respiratory induced deformation in the thorax and abdominal region. Such models relay on anatomical shape to correct intra-patient deformation fields extracted from 4D-CT data. An important application for liver RT is the argumentation of the liver segment in 4D-CT images with MRI information [8] to make lesions visible.

Deep learning-based image registration is here a very promising attempt, not only in terms of the short calculation time based on the GPU, but also because it provides the probability to enforce more complex requirements on the deformation field over classical regularization used in iterative DIR. A registration network (Reg-Net) is typically used to predict the deformation fields, which can be trained by

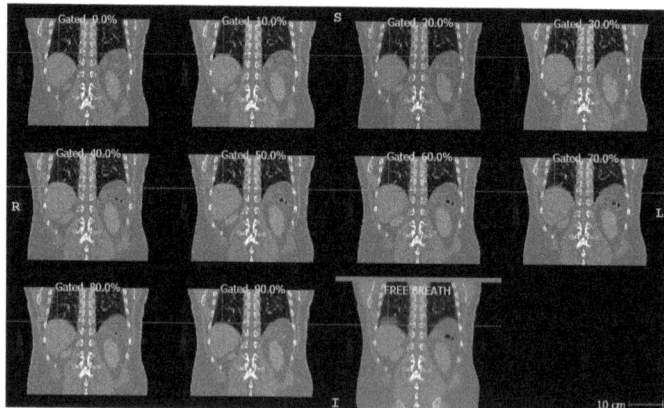

Fig. 1 A exemplar of 4D-CT, where ten phases in a respiration circle and one free breathing scan are revealed. The first five phases indicate the motion from the end of inhalation (EOI) to the end of the exhalation (EOE). The free breathing scan can be checked from the last image, which has many motion artifacts caused by the respiration

supervised [9], weakly supervised [10], or unsupervised manner [11]. In supervised manner, the deformation field are used as the ground-truth to train the network. For weakly supervised method, matching contours of delineated organs [10] are required in form of evaluating the overlap in the loss function. This is considered as a tremendous advantage because it enables for the implicit learning of anatomical constraints over relaying only on image contrast. More generally, the unsupervised training method is easily used in the application stage, as it facilitates the learning of Reg-Net without the need of expert delineation. Some works [12, 13] have shown the feasibility of Reg-Net in clinical applications. Wei *et al.* propose to use Reg-Net to perform MR-CT registration for liver tumor ablation guidance [12]. Yang *et al.* propose a patch-based Reg-Net to perform 4D-CT image registration [13].

In this paper, we propose to use a deep learning based registration network for 4D-CT images, and show its two applications. Instead of predicting the deformation fields by patch, we train our network to predict the deformation field in an end-to-end manner to avoid the deformation field fusion error. The registration network in our method is trained in an unsupervised manner without the need for the expert anno-tation. For further evaluating the performance in the application stage, we analysis the registration performance over different tissues in the abdominal and thoracic 4D-CT images. The experimental results reveal that our method can improve the tissue alignment. Based on Reg-Net, we perform the prediction of missing phase and the tracking of respiratory, which clearly shows the motion pattern in 4D-CT.

2 Methods

We use an unsupervised learning based Reg-Net to perform deformable intra-subject registration of any two phases in a 4D-CT image, as shown in Fig. 2. For training the Reg-Net in an unsupervised manner, the spatial transformer [14] is applied to warp the moving phase with the computed deformation field so as to compare the similarity between the warped phase and the fixed phase.

2.1 Training Loss

Here we use the unsupervised learning to train a network to perform deformation image registration over the 4D-CT images. We define the loss function:

$$\mathcal{L} = ||I_m(\phi) - I_f|| + \alpha|| \bigtriangledown \phi|| \tag{1}$$

where I_m, I_f represent the moving phase and the fixed phase, respectively. ϕ indicates the deformation field. α is a loss weight to balance the importance between two terms. In our implementation, we set it as 0.5. The first term is to compute the dissimilarity

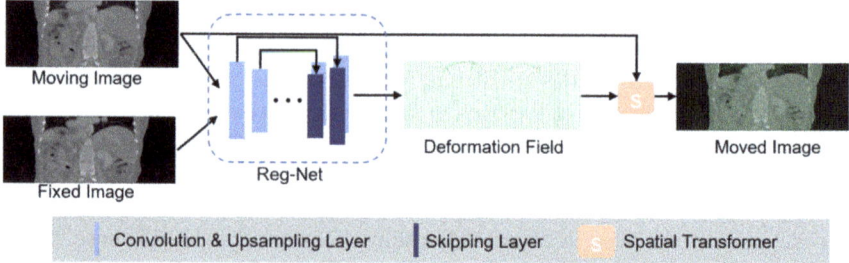

Fig. 2 The framework of the unsupervised deformation registration

between the moved phase $I_m(\phi)$ and the fixed phase I_f. The second term is to compute the smoothness of the deformation field ϕ. Therefore, such loss function drives the network to generate physically meaningful deformation fields, which is capable to warp the moving phase to be similar with the fixed phase, and keep the smoothing property.

2.2 Training Dataset

We train the network with two neighboring phases of one subject in each iteration, as shown in the Fig. 2. In each iteration at the training stage, one image pair consisting of the moving/fixed phases is input into the network. The output is the deformation field, with which the spatial transformer [14] can transform the moving phase to obtain the moved phase. In the inference stage, we can use the network in a regression manner. The deformation field can be computed at a very fast speed when two phases are given.

2.3 Network Framework

Here we use one 3D U-Net architecture akin to [11] as the registration network (as shown in Table 1), which inputs with two images and outputs one deformation field. The architecture can be checked from Fig. 2, where four down-sampling layers are used. The moving and fixed phases are concatenated as the input. The output is the deformation field, which is used to warp the moving phase.

Table 1 The layer configurations of Reg-Net. Conv3D represents the 3D convolution layer with the specified filter size, stride and number of filters. The Padding column indicates whether the Conv3D followed by zero padding to keep feature map size same. Skip connection is shown using Concat, which concatenates the previous nearest neighbor upsampled feature map with the corresponding Conv3D's feature map

Layer name	Filter size	Number of filter	Stride	Padding	Nonlinearity
Concat_1(Moving, Fixed)					
Conv3D_1	$3 \times 3 \times 3$	2	1	Y	LeakyReLU(0.2)
Conv3D_2	$3 \times 3 \times 3$	2	1	Y	LeakyReLU(0.2)
Conv3D_3	$3 \times 3 \times 3$	16	2	Y	LeakyReLU(0.2)
Conv3D_4	$3 \times 3 \times 3$	32	1	Y	LeakyReLU(0.2)
Conv3D_5	$3 \times 3 \times 3$	32	2	Y	LeakyReLU(0.2)
Conv3D_6	$3 \times 3 \times 3$	32	1	Y	LeakyReLU(0.2)
Conv3D_7	$3 \times 3 \times 3$	32	2	Y	LeakyReLU(0.2)
Conv3D_8	$3 \times 3 \times 3$	32	1	Y	LeakyReLU(0.2)
Conv3D_9	$3 \times 3 \times 3$	32	2	Y	LeakyReLU(0.2)
Conv3D_10	$3 \times 3 \times 3$	32	1	Y	LeakyReLU(0.2)
Conv3D_11	$3 \times 3 \times 3$	32	1	Y	LeakyReLU(0.2)
Upsampling3D_1		32	2		
Concat_2(Conv3D_8)		32+32			
Conv3D_12	$3 \times 3 \times 3$	32	1	Y	LeakyReLU(0.2)
Upsampling3D_2		32	2		
Concat_3(Conv3D_6)		32+32			
Conv3D_13	$3 \times 3 \times 3$	32	1	Y	LeakyReLU(0.2)
Upsampling3D_3		32	2		
Concat_4(Conv3D_4)		32+32			
Conv3D_14	$3 \times 3 \times 3$	32	1	Y	LeakyReLU(0.2)
Conv3D_15	$3 \times 3 \times 3$	16	1	Y	LeakyReLU(0.2)
Upsampling3D_4		16	2		
Concat_4(Conv3D_2)		2+16			
Conv3D_16	$3 \times 3 \times 3$	6	1	Y	LeakyReLU(0.2)
Conv3D_17	$3 \times 3 \times 3$	3	1	Y	Linear

3 Experiments and Results

Dataset and Pre-processing—Totally 15 4D-CT subjects are collected from the University of Colorado and Washington University School of Medicine, acquired on different CT scanners (Philips Brilliance 64, Philips Brilliance Big Bore, and Siemens Somatom Definition AS). 10 of them are used to train the Reg-Net, and 5 of them are used to test the accuracy. As each subject has 10 3D CT images corresponding to

10 phases, there are $10 \times 10 \times 9 = 900$ different 3D CT pairs in the training stage. And a random pair is input into Reg-Net at each iteration. Although the number of subjects is limited, the number of training pairs is sufficient to train a model to obtain acceptable performance when registering different phases in 4D-CT. Several phases of the testing subjects are annotated by clinical experts, including lungs, heart, stomach, liver, trachea, gross target volume (GTV), interal target volume (ITV), and so on. Among them, GTV and IVT are the main target regions in RT operation. The subject scans have various voxel spacing range from $0.976563 \times 0.976563 \times 3$ mm^3 to $1.17188 \times 1.17188 \times 3$ mm^3. And the scan sized from $512 \times 512 \times 73$ to $512 \times 512 \times 141$. For the pre-processing step, we normalize them to be fixed settings as below: (i) voxel spacing resampled as $1 \times 1 \times 3$ mm^3; (ii) crop and pad the input image size to be $256 \times 256 \times 96$.

Implementation—The network was implemented in Keras and trained on a single NVIDIA Tesla P40 GPU. At the training stage, two phases of one random subject are chosen to act as the input image pair. Totally, 60000 iterations are performed to train the network.

Evaluation Metrics—We computed the Dice similarity coefficient (DSC) and intersection over union (IoU) over two annotated organs or regions (V_1, V_2), which are given:

$$\text{DSC} = 2 \cdot \frac{V_1 \cap V_2}{V_1 + V_2} \tag{2}$$

$$\text{IoU} = \frac{V_1 \cap V_2}{V_1 \cup V_2} \tag{3}$$

The higher DSC and IoU indicates better registration performance. Also, such two metrics are of high interest in RT.

3.1 Registration

We performed intra-subject registration over the test dataset using Reg-Net, identity transformation, and NiftyReg. NiftyReg is chosen as the baseline method, concerning its superior performance in several applications [15]. It can be seen that our proposed methods can well align the lung boundary in 4D-CT, as shown in Fig. 3. For each subject in the testing dataset, randomly two scans were registered by our Reg-Net to obtain the deformation field. The annotation label is transformed by this deformation field. Concerning the motion may be different over different tissues, we compare the DSC and IoU over 12 organs or regions, as listed in Table 2.

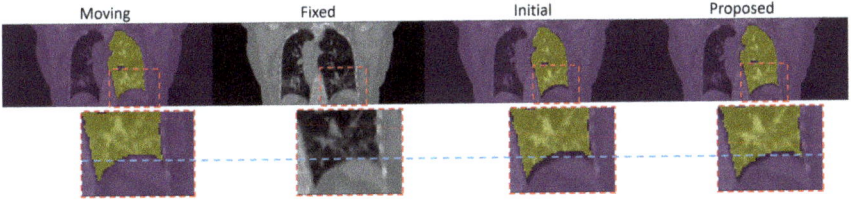

Fig. 3 The result of our method compared with initial identity transformation

Table 2 The mean±std DSC and IoU of initial annotation, and registration results by NiftyReg and our method over the testing dataset

	DSC (%)			IoU (%)		
	Initial	NiftyReg	Ours	Initial	NiftyReg	Ours
GTV	26.11±21.64	35.09±16.87	**49.31±14.22**	17.05±16.46	28.22±18.4	**33.99±13.49**
ITV	39.30±21.16	47.26±13.73	**58.51±12.99**	26.85±18.27	35.78±15.85	**42.63±13.92**
Carina	53.05±14.10	60.08±8.71	**62.36±11.17**	37.37±13.36	43.61±11.07	**46.24±11.57**
Esophagus	73.23±1.61	73.38±1.99	**74.61±1.74**	57.79±2.00	58.55±2.92	**59.53±2.21**
Heart	83.22±3.25	86.43±1.81	**87.07±2.69**	71.39±4.72	76.43±3.14	**77.20±4.18**
Left Bronchus	39.54±13.26	50.55±14.01	**58.27±14.85**	25.42±9.45	38.33±6.54	**42.63±14.56**
Right Bronchus	48.75±9.61	51.14±10.33	**54.97±9.44**	32.75±8.07	34.99±9.01	**38.48±8.90**
Liver	90.02±2.14	91.11±2.10	**92.10±1.62**	81.92±3.53	83.65±2.50	**85.39±2.79**
Left lung	92.04±3.02	93.71±2.83	**94.97±2.01**	85.40±5.17	88.97±4.81	**90.49±3.61**
Right lung	92.79±1.18	93.74±1.09	**94.71±0.93**	86.59±2.04	87.15±1.81	**89.96±1.68**
Rib	69.53±0.00	70.18±0.01	**71.02±0.00**	53.29±0.00	54.38±0.01	**55.06±0.00**
Stomach	72.18±6.06	68.87±5.89	**76.39±5.23**	56.81±7.25	58.99±6.70	**62.09±6.78**
Trachea	84.80±2.48	85.10±2.77	**86.71±2.74**	73.70±3.74	74.71±4.50	**76.64±4.30**

It can be observed that the DSC and IoU can be significantly improved by our method than identity transformation for each ROI, especially for the GTV and ITV, *e.g.,* p-values <0.05. For the organs in risk (heart, lung, liver, stomach and so on), the initial annotation transforming can obtain a high overlap. Even so, our method marginally improve the overlap ratio over these regions.

Furthermore, we analyse the correlation between the DSC of initial transforming and the DSC of our method, as shown in Fig. 4. We plot the figure to make the initial DSC as the x-axis, and DSC of Reg-Net as y-axis. It is shown that our method can improve the DSC over all the tissues in every testing pair. Especially for the ROIs with lower initial DSC, our method can improve the DSC by a significant value.

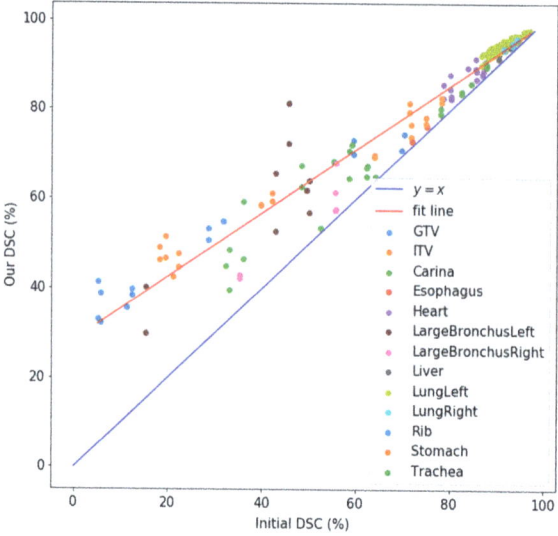

Fig. 4 The Dice similarity coefficient (DSC) correlation between our method and the initial identity transformed result over different tissues or organs. The fit line is computed from the scatter by least squares method

3.2 Interpolation

Given two input phases (I_0, I_1), the deformation field ϕ is obtained by our trained Reg-Net. The intermediate phase can be obtained by the warping operation using weighted deformation field ϕ_α.

$$I_\alpha = I_0(\phi_\alpha) \tag{4}$$

For example, we can interpolate the middle phase between the EOI and EOE by setting $\alpha = 0.6$, which is larger than expected 0.5 due to registration error (shown in Fig. 5a).

3.3 Motion Tracking

Based on the our proposed registration method, we also perform motion tracking. In our experiment, we choose the EOI phase to register with the left 9 phases in 4D-CT images. We place a sampling point over the lung bottom boundary to show its displacement, which is computed from the obtained deformation fields, as shown in Fig. 5b.

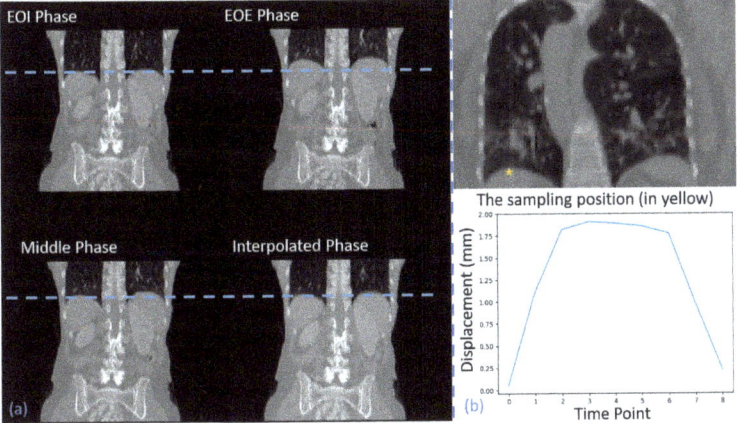

Fig. 5 a The interpolation exemplar. Given the EOI and EOE phases, the middle phase can be obtained by warping the weighted deformation field. **b** The motion tracking (bottom) of yellow point (top) in lung bottom boundary

4   Conclusion

An unsupervised registration network is proposed to align the 4D-CT image. Experimental results show that our method can align any two phases in the 4D-CT images. Furthermore, we analysis the results of registration network over the different tissues or organs. Our proposed method can obtain improvement over all of the ROIs, especially on the volume of target in radiation therapy. Based on the proposed registration method, the missed phase interpolation and the motion tracking can be well performed. In future, more comparison experiments need to be conducted, and more metrics (*i.e.*, distance between annotated organ surface) should be involved to comprehensively evaluate the performance of 4D-CT registration performance by Reg-Net.

References

1. Brandner, E. D., Wu, A., Chen, H., Heron, D., Kalnicki, S., Komanduri, K., Gerszten, K., Burton, S., Ahmed, I., & Shou, Z. (2006). Abdominal organ motion measured using 4d ct. *International Journal of Radiation Oncology* Biology* Physics, 65*(2), 554–560.
2. Tai, A., Liang, Z., Erickson, B., & Allen Li, X. (2013). Management of respiration-induced motion with 4-dimensional computed tomography (4dct) for pancreas irradiation. *International Journal of Radiation Oncology Biology Physics, 86*(5), 908–913.
3. D'Souza, Warren D., Nazareth, Daryl P., Zhang, Bin, Deyoung, Chad, Suntharalingam, Mohan, Kwok, Young, et al. (2007). The use of gated and 4d ct imaging in planning for stereotactic body radiation therapy. *Medical Dosimetry, 32*(2), 92–101.
4. Keall, Paul. (2004). 4-dimensional computed tomography imaging and treatment planning. *Seminars in Radiation Oncology, 14*(1), 81–90.

5. Gupta, V., Wang, Y., Romero, A., Myronenko, A., Jordan, P., Maurer, C., Heijmen, B., & Hoogeman, M. (2018). Fast and robust adaptation of organs-at-risk delineations from planning scans to match daily anatomy in pre-treatment scans for online-adaptive radiotherapy of abdominal tumors. *Radiotherapy and Oncology*, *127*(2), 332–338.

6. McClelland, J. R., Hawkes, D. J., Schaeffter, T., & King, A. P. (2013). Respiratory motion models: A review. *Medical Image Analysis*, *17*(1), 19–42.

7. Towards a generic respiratory motion model for 4D CT imaging of the thorax. (2009). *IEEE Nuclear Science Symposium Conference Record*, 3975–3979.

8. Harris, Wendy, Wang, Chunhao, Yin, Fang-Fang., Cai, Jing, & Ren, Lei. (2018). A Novel method to generate on-board 4D MRI using prior 4D MRI and on-board kV projections from a conventional LINAC for target localization in liver SBRT. *Medical Physics*, *45*(7), 3238–3245.

9. Eppenhof, K. A. J., & Pluim, J. P. W. (2019). Pulmonary CT registration through supervised learning with convolutional neural networks. *IEEE Transactions on Medical Imaging*, *38*(5), 1097–1105.

10. Dalca, Adrian V., Balakrishnan, Guha, Guttag, John V., & Sabuncu, Mert R. (2019). Unsupervised learning of probabilistic diffeomorphic registration for images and surfaces. *Medical Image Analysis*, *57*, 226–236.

11. Balakrishnan, G., Zhao, A., Sabuncu, M. R., Guttag, J., & Dalca, A. V. (2019). Voxelmorph: A learning framework for deformable medical image registration. *IEEE transactions on medical imaging*, *38*(8), 1788–1800.

12. Wei, D., Ahmad, S., Huo, J., Peng, W., Ge, Y., Xue, Z., Yap, P. T., Li, W., Shen, D., & Wang, Q. (2019). Synthesis and inpainting-based mr-ct registration for image-guided thermal ablation of liver tumors. In *22nd International Conference on Medical Image Computing and Computer-Assisted Intervention, MICCAI 2019*, pp. 512–520.

13. Lei, Y., Fu, Y., Harms, J., Wang, T., Curran, W. J., Liu, T., Higgins, K., & Yang, X. (2019). 4d-ct deformable image registration using an unsupervised deep convolutional neural network. *Workshop on Artificial Intelligence in Radiation Therapy*, pp. 26–33.

14. Jaderberg, M., Simonyan, K., Zisserman, A., Kavukcuoglu, K. (2015). Spatial transformer networks. In *NIPS'15 Proceedings of the 28th International Conference on Neural Information Processing Systems*, pp. 2017–2025.

15. Rueckert, D., Sonoda, L. I., Hayes, C., Hill, D. L. G., Leach, M. O., & Hawkes, D. J. (1999). Nonrigid registration using free-form deformations: Application to breast MR images. *IEEE Transactions on Medical Imaging*, *18*(8), 712–721.

3D Reconstruction of Patient-Specific Carotid Artery Geometry Using Clinical Ultrasound Imaging

Tijana Djukic, Branko Arsic, Igor Koncar, and Nenad Filipovic

Abstract One of the techniques used to diagnose carotid artery disease is the ultrasound (US) examination. The initiation and development of vascular diseases depends also on the flow conditions in the artery. Additional parameters that cannot be directly measured can be obtained by performing numerical simulations using patient-specific geometry. In this study, the Finite Element Method (FEM) is used to analyze the distribution of relevant blood flow characteristics. Images obtained from the US examinations are used to adapt the generalized carotid bifurcation model to the specific patient. The approach presented in this study combines the deep learning approach for the image segmentation and automated 3D reconstruction method to create a semi-generic geometrical model of the carotid artery that is adapted to the specific patient, using data obtained from only several US images. The presented methodology enables efficient segmentation, extraction of the morphological parameters and creation of 3D meshed volume models that can be also used for the further computational simulations of blood flow.

T. Djukic (✉)
Institute for Information Technologies, University of Kragujevac, Kragujevac, Serbia
e-mail: tijana@kg.ac.rs

T. Djukic · B. Arsic
Bioengineering Research and Development Center, BioIRC, Kragujevac, Serbia

B. Arsic
Faculty of Science, University of Kragujevac, Kragujevac, Serbia

I. Koncar
Clinic for Vascular and Endovascular Surgery, Serbian Clinical Centre, Belgrade, Serbia

N. Filipovic
Faculty of Engineering, University of Kragujevac, Kragujevac, Serbia

© The Author(s), under exclusive license to Springer Nature Switzerland AG 2021
K. Miller et al. (eds.), *Computational Biomechanics for Medicine*,
https://doi.org/10.1007/978-3-030-70123-9_6

1   Introduction

One of the diseases in human cardiovascular system is carotid artery stenosis (CAS). Early detection of this disease is very important because if it is not adequately treated, it may potentially have deteriorating consequences, such as a debilitating stroke. These serious accidents occur when atherosclerotic plaques in the arteries suddenly rupture, leading to the obstruction of the blood flow to the heart or to the brain. There are several examination techniques that can be applied to analyze the state of patient's carotid artery, including both 3D imaging techniques like computed tomography (CT) and magnetic resonance imaging (MRI) [1], as well as ultrasound (US) examination [2–4]. US technique is accurate, noninvasive and inexpensive and is hence one of the first methods that is applied in diagnosing carotid artery disease. The US combines two procedures: traditional B-mode (gray-scale) ultrasound and color-Doppler ultrasound. Using the first procedure, images of the vessels in question at rest are created from the reflected sound waves. The second procedure provides information about the motion of the blood and enables the visualization of blood flow and measurement of flow velocity. Both these procedures produce two-dimensional (2D) cross-sectional images.

Deep learning is a promising machine learning tool for the automatic classification and interpretation of medical image data. This technique has been applied extensively in several medical imaging applications, such as for brain, lung, and breast imaging. Its application to arterial structures such as the carotid artery and for noisy image data as those found in US images of the carotid bifurcation has not been reported yet [5]. Analysis of a carotid US images requires segmentation of the vessel wall, lumen, and plaque of the carotid artery. Convolutional neural networks are one of the most common tools used in image segmentation. In [6], the U-Net convolutional neural network was used for lumen segmentation from US images of the entire carotid system.

After obtaining the US images, additional analysis can be performed. Since the initiation and development of vascular diseases depends also on the flow conditions in the artery, it is useful to analyze the distribution of relevant blood flow characteristics. The US examination can be used to measure wall thickness and blood velocities in patients. Additional parameters that cannot be directly measured can be obtained using numerical simulations, more precisely computational fluid dynamics (CFD) methods. In this study, the Finite Element Method (FEM) is used to analyze blood velocity, pressure and wall-shear stress distribution [7–9].

The simulations can be performed using idealized carotid bifurcation, but in order to determine the correct patient-specific diagnosis that also considers the individual anatomy of the particular patient, it is necessary to perform simulations using a patient-specific geometry. In this study, images obtained using the US examinations are used to adapt the generalized carotid bifurcation model to the specific patient.

The paper is organized as follows: The applied methods and the used clinical data are discussed in Sect. 2. The results of the training of the neural network and the blood flow simulation results for a specific patient are presented in Sect. 3. Section 4 discusses relevant work in literature and concludes the paper.

2 Materials and Methods

In this Section, the details of the applied techniques will be given.

2.1 Clinical Data

In order to develop and validate the tool for 3D reconstruction of carotid artery based on the US images as input, a dataset of original and annotated US images obtained in the Serbian Clinical Centre, was used. The annotated images were used for processing and training, while the original images were used for the validation. The dataset consisted of 108 patients who underwent the US examination (baseline time point). For each patient the common carotid artery, the branches and carotid bifurcation in transversal and longitudinal projections are captured, which included overall 876 images in the data set. The examination was performed in B-mode and Color-Doppler mode. All imaging data were anonymized respecting the data protection and safety.

2.2 Extraction of Vessel Lumen from US Images Using Deep Learning

The preprocessing of US dataset included following actions:

- Annotation of carotid lumen area
- Resizing/Cropping of US images to 512×512 pixels
- Classification of longitudinal and transversal US images
- Classification of B-mode and Color-Doppler mode images.

The automatic carotid artery segmentation is done using U-Net [10] based deep convolutional networks. U-Net is a convolutional neural network for image segmentation with the most important application being in segmentation of medical images. It is based on encoder-decoder model. The variant of U-Net in this study is slightly modified from the original. It has two additional blocks in both encoder and decoder. Each block in encoder has two convolutional layers with 3×3 filters, followed by 2×2 max pooling. In each decoder block, 2×2 up-convolution and skip connection are followed by three more convolutional layers with 3×3 filters, and the last decoder block produces the segmentation mask with 1×1 convolution and sigmoid activation function. All convolutional layers are padded so that the resulting activation map preserves the same height and width. In this way, the resulting segmentation map has the same resolution as the input image. Also, the variant of U-Net applied in this study uses batch normalization after each convolutional layer which proves to work a lot better on our data than the original U-Net model [11]. All batch normalization layers are followed by a ReLU activation.

The model is trained with a combination of binary cross-entropy and soft dice coefficient as a loss function, which is expressed as:

$$Loss = binary\_crossentropy(y_{true}, y_{pred}) + 1 - dice\_coeff(y_{true}, y_{pred}) \quad (1)$$

where y_{pred} and y_{true} denote the flattened predicted probabilities and the flattened ground truths of the image. Soft Dice coefficient loss is described in [12, 13].

From the entire dataset explained in Sect. 2.1, all images corresponding to the patients are randomly divided into training, validation and testing sets by a ratio of 8:1:1 at the carotid artery level (either for the left or for the right arterial model), such that a total of 700 images have been taken out for training purposes and the remaining is used for validation and testing.

2.3   3D Reconstruction of the Carotid Artery

The 3D reconstruction of patient-specific carotid artery is performed using the available clinical imaging data for the particular patient. One of the main problems with the used patient data set is the limited number of 2D transversal cut that is available. In order to overcome the problem with the missing cuts, the generalized model was used as the basis, that is then improved with the available data, by mapping the available transversal cuts at specific positions in the generalized model. This process is illustrated in Fig. 1. The generalized model was defined according to data presented in literature [14, 15]. The transversal cut of the common carotid artery (CCA—annotated by the B line in Fig. 1) and the external carotid artery (ECA - the lower segment annotated by the C line in Fig. 1) is used to define the shapes of all cross-sections of the carotid artery in these segments. The longitudinal cut of the internal carotid artery (ICA—annotated by the A segment in Fig. 1) is used to extract the centerline of the ICA and the diameters in this segment, while the transversal cuts of the ICA (the upper segment annotated by the C line and the cross-section annotated by the D line in Fig. 1) are also used to more accurately define these particular cross-sections. The lengths of the branches are defined according to the information obtained from the cuts. Length of the ECA was taken to be half of the length of the ICA that was extracted from the longitudinal cut. The length of the CCA was set to be 1.2 times higher than the diameter of the CCA that was extracted from the appropriate transversal cut.

The 3D reconstruction process is performed using the algorithm similar to the one presented in literature [16] and the overview of this process is shown in Fig. 2. First, the extracted segments are obtained using the procedure explained in Sect. 2.2 (Fig. 2A). Then, the smoothing of the obtained curves is performed (Fig. 2B), by converting them to nonuniform B-spline curves. After that, the cross-sections along the centerline are defined (Fig. 2C) and finally, the 3D FE mesh is generated (Fig. 2D). Details of these steps are given in the sequel.

Fig. 1 The adaptation of the generalized model of the carotid artery using US images obtained for a specific patient

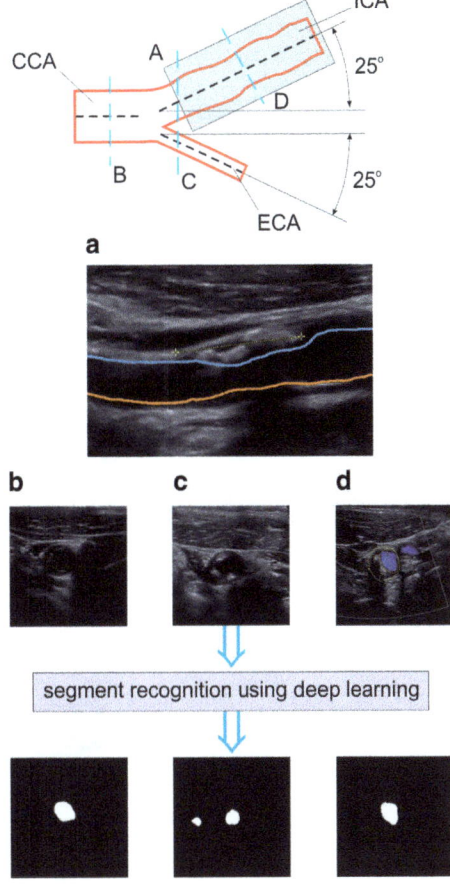

Individual branches (CCA, ICA and ECA) are modeled as tube-like surface. For each branch, a parameterized centreline $\overset{\rho_i}{c}(t)$ is defined (where index i represents the branch (CCA is denoted by 1, ICA by 2 and ECA by 3):

$$\overset{\rho_i}{c}(t) = \sum_{i=1}^{q} \mathbf{x}_i N_{i,k}(t) \tag{2}$$

where $0 \leq i \leq 1$ and $2 \leq k \leq i+1$. In Eq. (1) the control points of the centerline are defined with $\mathbf{x}_i \mathbf{x}_i$, and $N_{i,k}$ denotes the k-th order basis functions, that are calculated according to the Cox–de Boor recursive algorithm [17].

The trihedron of the centerline $\overset{\rho_i}{c}(t)$ is defined using the Frenet–Serret formulas [17]. The trihedron is defined via the curve's tangent $\overset{\rho_i}{T}(t)$, normal $\overset{\rho_i}{N}(t)$ and binormal

Fig. 2 The process of 3D
reconstruction; **a**—segment
extracted from US image;
b—nonuniform B-spline;
c—parameterized centerline
with cross-sections; **d**—3D
FE mesh

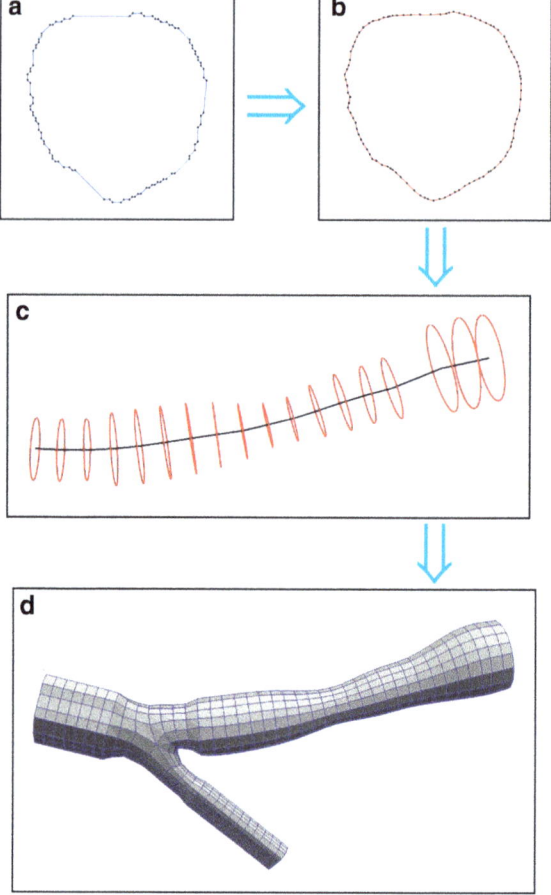

$\overset{\rho_i}{B}(t)$. The tube-like surface for each branch is defined with cross-sections positioned along the parameterized centerline. The cross-sections are either circles with extracted diameters, or the extracted segments. In both cases, the points of the cross-sections are projected onto the trihedron normal-binormal plane of each centerline to obtain the patches of the surface. In order to ensure the regularity of points and to prevent twisting of individual patches, all points were converted to polar coordinates in normal-binormal plane and sorted in circular direction. These patches in 3D space are used for NURBS surface representation that can be defined as:

$$\overset{\rho_i}{S}(u, v) = \sum_{i=1}^{q} \sum_{j=1}^{w} \mathbf{B}_{i,j} N_{i,k}(u) M_{i,k}(v) \tag{3}$$

where $0 \le u \le 1$ and $0 \le v \le 1$. In Eq. (3) the homogenous points of the control net polygon are defined with $\mathbf{B}_{i,j}$, and $N_{i,j}$ and $M_{i,j}$ denote the k-th and l-th order basis functions, respectively. The basis functions are calculated using the same approach used in Eq. (2), according to the Cox–de Boor recursive algorithm [17].

Using the parameterized vessel centerline $\overset{\rho_i}{c}(t)$ and surface $\overset{\rho_i}{S}(u, v)$ it is possible to perform the discretization of the vessel. The introduction of the parameterization enabled to control the meshing process by defining the density of nodes in the mesh in both longitudinal (u) and circular (v) directions. Details of the meshing procedure are described in [16]. The main idea is to decompose the branches of the carotid artery into connected hexahedra elements, which were composed of quadrilateral patches (that are actually the vessel cross-sections). Finally, the individual branches of the carotid artery are then connected into a bifurcation, following the procedure described in [16].

3 Results

3.1 Image Segmentation Results for the Lumen

In order to present the results of the neural network training, the binary classification task for image segmentation was considered. Four common classification metrics are considered for quantitative evaluation, including precision (P), recall (R), F1 score and Dice coefficient [18]. P metric effectively describes the purity of the positive detections in the image, relative to the ground truth, while R metric effectively describes the completeness of the positive predictions in the image, relative to the ground truth. The F1 score combines the P and R metrics to illustrate if the model represents an optimal blend of precision and recall, and the Dice coefficient is a common metric that measures the overlap between the predicted and the ground truth. The results for the test set are shown in Table 1. The method used in this study is compared with thresholding technique and FCN-8 s model with VGG16 as a backbone classifier [19]. Similar to U-Net, FCN model proved to work better with batch normalization. However, FCN has almost twice as many trainable parameters than U-Net network, so the U-Net can be trained faster and is more memory efficient.

Table 1 U-Net results on test dataset

Precision	Recall	F1-score	Dice coefficient
0.90	0.92	0.91	0.90

3.2 Blood Flow Through the 3D Reconstructed Carotid Artery

The results of the 3D reconstruction and blood flow simulations are presented in this Section. Clinical data for a patient that belonged to the testing set is used for the 3D simulations. The reconstruction is performed using the procedure described in Sect. 2.3, while the blood flow simulations were performed using the software PakF [8, 9]. The fluid is observed as a Newtonian fluid and its characteristics were defined according to the literature [8, 14]. The fluid density is set to 1.05 g/cm^3 and the kinematic viscosity is set to 0.035 cm^2/s. The initial condition is defined using the inlet velocity (at the inlet of CCA) that was set to be equal to 96 cm/s, for all cases. The outflow boundary condition is defined at the outlet branches (ICA and ECA).

The fluid flow results are shown in Fig. 3, which includes the pressure distribution (Fig. 3a), wall shear stress (WSS) distribution (Fig. 3b) and velocity streamlines (Fig. 3c).

4 Discussion and Conclusion

There are many papers in literature that use B-mode US images to analyze the state of the patient's carotid arteries. Kristensen et al. [20] performed US examinations on 1390 patients in a clinical study, which provided fast diagnostics and enabled better clinical decision-making. The US imaging was also used to monitor the progression of atherosclerotic plaque [21]. In combination with the Color-Doppler US, it is possible to obtain quantitative estimates of luminal narrowing and other flow measurements. Visualizing and analyzing the morphological structure of carotid bifurcations are important for understanding the etiology of carotid atherosclerosis, which is a major cause of stroke and transient ischemic attack. However, conventional 2D ultrasound imaging has technical limitations in observing the complicated 3D shapes and asymmetric vasodilation of bifurcations, since it is not possible to obtain a composite view of the vessel wall, lumen and plaque. This problem was solved with 3D US imaging that was successfully used in literature to pre-surgically evaluate the carotid stenosis [22].

The three-dimensional (3D) visualization of the carotid artery from the 2D US images would also be very useful. This would be possible if a series of cross-sectional B-mode US images is acquired. Rosenfield et al. [2] used B-mode US images to perform 3D reconstruction of the carotid bifurcation from its inward flow. Authors of a more recent paper [23] also performed 3D reconstruction from 2D transversal US images.

Another limitation of US imaging is that it requires skills, expertise and knowledge from the sonographer when making data acquisition. It also influences the quality of the images. Therefore, data taken by different sonographer has the possibility of

Fig. 3 Results of fluid flow simulation; **a**—pressure distribution; scale bar on the right is in Pa; **b**—WSS distribution; scale bar on the right is in Pa; **c**—velocity streamlines; scale bar on the right in m/s

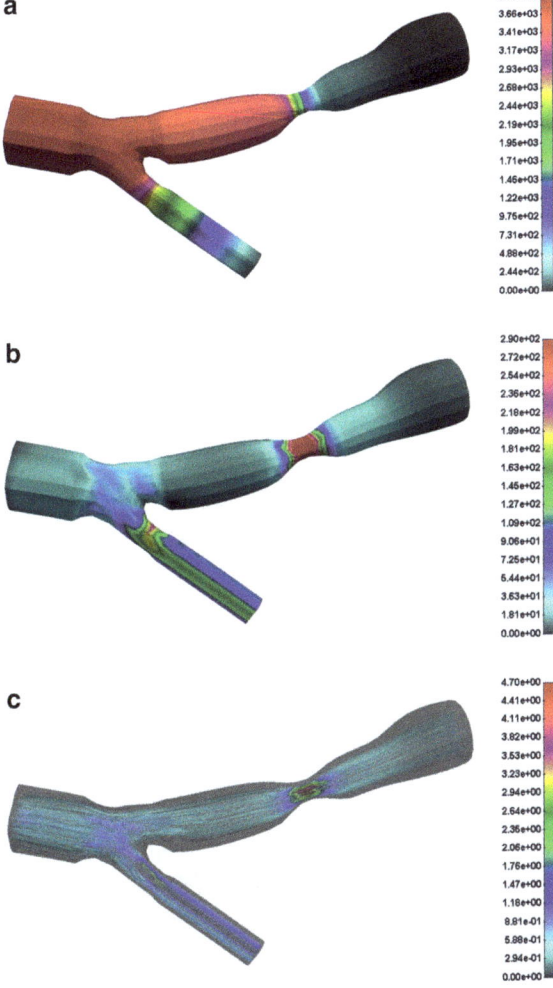

producing different images. The difference in image results can affect the process of locating and segmenting CA areas. Automatic determination of the location of the CA area can be used to avoid the subjectivity of the sonographer. However, there are not many methods proposed in literature that automatically reconstruct carotid stenosis and/or bifurcation with vessel course alterations from US images. The proposed methodology includes training of a deep learning model (U-net) for the image segmentation phase and a computer-based automated 3D reconstruction method that is capable of generating a semi-generic geometrical model of the carotid artery that is adapted to the specific patient, using data obtained from only several transversal US images.

The presented approach and methodology which combines the data mining and 3D reconstruction of carotid artery enables efficient segmentation, extraction of the morphological parameters and creation of 3D meshed volume models that can be also used for the further computational simulations of blood flow.

Acknowledgements The research presented in this study was part of the project that has received funding from the European Union's Horizon 2020 research and innovation programme under grant agreement No. 755320-2 - TAXINOMISIS. This article reflects only the author's view. The Commission is not responsible for any use that may be made of the information it contains. The research is also supported by the Ministry of Education, Science and Technological Development of the Republic of Serbia (project numbers III41007 and ON174028).

References

1. Sadat, U., Teng, Z., Young, V. E. et al. (2010). Three-dimensional volumetric analysis of atherosclerotic plaques: A magnetic resonance imaging-based study of patients with moderate stenosis carotid artery disease. *The International Journal of Cardiovascular Imaging*. https://doi.org/10.1007/s10554-010-9648-6.
2. Rosenfield, K., Losordo, D. W., Ramaswamy, K., et al. (1991). Three-dimensional reconstruction of human coronary and peripheral arteries from images recorded during two-dimensional intravascular ultrasound examination. *Circulation*. https://doi.org/10.1161/01.CIR.84.5.1938
3. Landry, A., Ainsworth, C., Blake, C., et al. (2007). Manual planimetric measurement of carotid plaque volume using three-dimensional ultrasound imaging. *Med. Phys., 10*(1118/1), 2715487.
4. Chiu, B., Beletsky, V., & Spence, J. D. et al. (2009). Analysis of carotid lumen surface morphology using three-dimensional ultrasound imaging. *Physics in Medicine and Biology*. https://doi.org/10.1088/0031-9155/54/5/004.
5. Lekadir, K., Galimzianova, A., & Betriu, À. et al. (2017). A convolutional neural network for automatic characterization of plaque composition in carotid ultrasound. *IEEE Journal of Biomedical and Health Informatics*. https://doi.org/10.1109/JBHI.2016.2631401.
6. Xie, M., Li, Y., Xue, Y. et al. (2019). Vessel lumen segmentation in internal carotid artery ultrasounds with deep convolutional neural networks. In *IEEE International Conference on Bioinformatics and Biomedicine (BIBM)*, San Diego, CA, USA, 2019, (pp. 2393–2398). https://doi.org/10.1109/BIBM47256.2019.8982980.
7. Parodi, O., Exarchos, T., & Marraccini, P. et al. (2012). Patient-specific prediction of coronary plaque growth from CTA angiography: A multiscale model for plaque formation and progression. *IEEE Transactions on Information Technology in Biomedicine*. https://doi.org/10.1109/TITB.2012.2201732.
8. Filipovic, N., Rosic, M., & Tanaskovic, I. et al. (2012). ARTreat project: Three-dimensional numerical simulation of plaque formation and development in the arteries. *IEEE Transactions on Information Technology in Biomedicine*. https://doi.org/10.1109/TITB.2011.2168418.
9. Filipovic, N., Teng, Z., & Radovic, M. et al. (2013). Computer simulation of three-dimensional plaque formation and progression in the carotid artery. *Medical Biological Engineering & Computing*. https://doi.org/10.1007/s11517-012-1031-4.
10. Ronneberger, O., Fischer, P., & Brox, T. (2015). U-net: Convolutional networks for biomedical image segmentation. In *Lecture Notes in Computer Science (including subseries Lecture Notes in Artificial Intelligence and Lecture Notes in Bioinformatics*. https://doi.org/10.1007/978-3-319-24574-4_28.
11. Zhou, X. Y., & Yang, G. Z. (2019). Normalization in training U-Net for 2-D biomedical semantic segmentation. *IEEE Robotics and Automation Letters*. https://doi.org/10.1109/LRA.2019.2896518.
12. Anbeek, P., Vincken, K. L., Van Bochove, G. S., et al. (2005). Probabilistic segmentation of brain tissue in MR imaging. *NeuroImage*. https://doi.org/10.1016/j.neuroimage.2005.05.046

13. Chang, H. H., Zhuang, A. H., Valentino, D. J., et al. (2009). Performance measure charac-
    terization for evaluating neuroimage segmentation algorithms. *NeuroImage*. https://doi.org/10.
    1016/j.neuroimage.2009.03.068
14. Perktold, K., Resch, M., & Peter, R. O. (1991). Three-dimensional numerical analysis of
    pulsatile flow and wall shear stress in the carotid artery bifurcation. *Journal of Biomechanics*.
    https://doi.org/10.1016/0021-9290(91)90029-M
15. Perktold, K., Peter, R. O., Resch, M., et al. (1991). Pulsatile non-newtonian blood flow in
    three-dimensional carotid bifurcation models: A numerical study of flow phenomena under
    different bifurcation angles. *Journal of Biomedical Engineering*. https://doi.org/10.1016/0141-
    5425(91)90100-L
16. Vukicevic, A. M., Çimen, S., & Jagic, N. et al. (2018). Three-dimensional reconstruction
    and NURBS-based structured meshing of coronary arteries from the conventional X-ray
    angiography projection images. Scientific Reports. https://doi.org/10.1038/s41598-018-194
    40-9.
17. Vukicevic, A. M., Stepanovic, N. M., Jovicic, G. R. et al. (2014). Computer methods for
    follow-up study of hemodynamic and disease progression in the stented coronary artery by
    fusing IVUS and X-ray angiography. *Medical Biological Engineering and Computing*. https://
    doi.org/10.1007/s11517-014-1155-9.
18. Hossin, M., & Sulaiman, M. N. (2015). *A review on evaluation metrics for data classification
    evaluations*. https://doi.org/10.5121/ijdkp.2015.5201
19. Long, J., Shelhamer, E., & Darrell, T. (2015). Fully convolutional networks for semantic
    segmentation. *Proceedings of the IEEE Computer Society Conference on Computer Vision
    and Pattern Recognition*. https://doi.org/10.1109/CVPR.2015.7298965
20. Kristensen, T., Hovind, P., Iversen, H. K., et al. (2018). Screening with doppler ultrasound for
    carotid artery stenosis in patients with stroke or transient ischaemic attack. *Clinical Physiology
    and Functional Imaging, 38*, 617–621.
21. Mallett C, House A A, Spence J D et al (2009) Longitudinal ultrasound evaluation of carotid
    atherosclerosis in one, two and three dimensions. *Ultrasound Medicine and Biology*. https://
    doi.org/10.1016/j.ultrasmedbio.2008.09.008.
22. Pfister, K., Rennert, J., & Greiner, B. et al. (2009). Pre-surgical evaluation of ICA-stenosis
    using 3D power doppler, 3D color coded doppler sonography, 3D B-flow and contrast
    enhanced B-flow in correlation to CTA/MRA: First clinical results. *Clinical Hemorheology
    and Microcirculation*. https://doi.org/10.3233/CH-2009-1161.
23. Yeom, E., Nam, K. H., Jin, C., et al. (2014). 3D reconstruction of a carotid bifurcation from
    2D transversal ultrasound images. *Ultrasonics*. https://doi.org/10.1016/j.ultras.2014.06.002

Evaluation of the Agreement Between Ultrasound-Based and Bi-Planar X-Ray Radiography-Based Assessment of the Geometrical Features of the Ischial Tuberosity in the Context of the Prevention of Seating-Related Pressure Injury

A. Berriot, N. Fougeron, X. Bonnet, H. Pillet, and P. Y. Rohan

Abstract The proper management of the local mechanical environment within soft tissues is a key challenge central the prevention of Pressure Ulcers (PUs). Magnetic Resonance (MR) imaging is the preferred imaging modality to measure geometrical features associated with PUs. It is a very time-consuming method and it represents a major barrier to the clinical translation of risk assessment tools. There is a growing enthusiasm of the community for the use of B-mode ultrasound imaging as a practical, alternative technology suitable for bedside or outpatient clinic use. The objective was to evaluate the agreement between US-derived measurements and bi-planar X-ray radiography-derived measurements of geometrical features of the Ischial Tuberosity in a realistic loaded sitting position in healthy volunteers. The reproducibility of the US-based assessment of radii of curvature, evaluated in a subset of 4 subjects using the ISO 5725–2 framework was 1.7 mm and 1.3 mm in the in the frontal and sagittal plane respectively (95% CI = 3.5 mm and = 2.6 mm respectively). Out of the 13 subjects included, the ischial tuberosity border was visible on the US image of 7 healthy subjects only. The mean of differences computed on the 7 subjects using Bland–Altman plots were + 3.3 mm and -5.7 mm in the frontal and sagittal planes respectively. The corresponding 95% CI in the frontal and sagittal planes were respectively 1.8 mm and 3.7 mm. These differences however were not statistically significant (Wilcoxon signed-rank test). More effort is needed to establish and standardise optimal measurement procedures and test protocols for the assessment of geometrical features of the IT using US.

Keywords Pressure ulcer · Soft tissue injury risk · Reproducibility · Ultrasound imaging · Bi-planar X-Ray radiography

A. Berriot · N. Fougeron · X. Bonnet · H. Pillet · P. Y. Rohan (✉)
Institut de Biomécanique Humaine Georges Charpark, Arts et Metiers Institute of Technology, 151 bvd de l'Hôpital, 75013 Paris, France
e-mail: Pierre-Yves.ROHAN@ensam.eu

1   Introduction

A Pressure Ulcer (PU) is defined as "*a localized injury to the skin and underlying soft tissue, usually over a bony prominence, caused by sustained pressure, shear or a combination of these*" (NPUAP, EPUAP, PPPIA 2014). It is a complication primarily related to the care and treatment of individuals who have difficulty moving or changing positions: for example, those people with a disability and the elderly [6]. It appears in situations where excessive mechanical loads are applied to the skin, such as, for example, during mechanical interaction between a person and support surfaces (hospital bed) or medical devices (Manual Wheelchair, prosthetic socket, exoskeleton, etc.). PU prevention remains a major health challenge for Europe due to the human and financial cost of prolonged hospitalization, reduced quality of life, loss of autonomy and social isolation. However, current risk assessment tools do not allow for a correct identification of the risks nor putting an effective prevention in place [5].

In the literature, a lot of research has sought to explain soft tissue injury risk factors in terms of the local mechanical environment (i.e. internal stresses and strains that satisfy mechanical equilibrium) within soft tissue [28]. In particular, it has been shown by combining an animal model of Deep Tissue Injury with computational modeling that direct deformation damage was only apparent when a certain strain threshold was exceeded [4]. In humans, computational modelling of load-bearing soft tissue has shown that bony prominences induce substantial stress concentrations, which explains why these areas are vulnerable to pressure ulcers [18]. Laboratory and animal studies propose several aetiological mechanisms by which stress and internal strain interact with damage thresholds to result in pressure ulcer development including localized ischaemia [20], reperfusion injury [15], impaired lymphatic drainage [13] and sustained cell deformation [3, 11].

Based on the rationale that elucidating the relationship between external loads and internal local stresses and strains within loaded soft tissues has the potential of improving the management and prevention of PUs, several Finite Element (FE) models of the buttock have been proposed in the literature based on MRI or CT scan data [2, 16, 17, 21, 30]. Because of the limited availability of MRI or CT-Scan systems and of the long segmentation time associated with the creation of full 3D subject specific FE models from these imaging systems, all the studies in the literature included the data of only one individual [2, 23]. As far as the authors are aware of, the only attempts to account for this variability were limited to (i) semi-3D modelling (N = 6 able-bodied volunteers in [17]; N = 12 in [18]; and N = 6 in [19]) and (ii) one attempt at 3D modelling (N = 6 in [27]). In addition, the representation of a realistic unloaded sitting position is jeopardized by the experimental limitations of MRIs and CT-scans: Long acquisition times of MR imaging prevent a prolonged unloaded sitting configuration without resorting to devices [2] while the confinement of the scanner limits the acquisition to the lying position only [23].

Recent developments in 3D reconstruction techniques from low dose calibrated bi-planar X-ray imaging (EOS imaging, Paris, France) provide a promising alternative tool for patient-specific validated 3D modelling of the pelvis [12]. Moreover, unlike CT scanner or MRI systems where the patient is in a supine position, this technique provide bi-planar images of the subject in a weight-bearing position.

In a previous study, and as an alternative to MRI-based/CT-scan-based assessment, a methodology combining low-dose biplanar X-ray images (EOS imaging, Paris, France), B-mode ultrasound images and optical scanner acquisitions in a non-weight-bearing sitting posture has been proposed for the fast generation of patient-specific FE models of the buttock and applied to 6 healthy subjects [25].To investigate the ability of a local model of the region beneath the ischium to capture the internal response of the buttock soft tissues predicted by a complete 3D FE model from a limited number of parameters, a simplified model was developed based on data compatible with daily clinical routine [24]. This study highlighted that the local mechanical environment within soft tissues is very sensitive (in the statistical sense i.e. variance-based sensitivity analysis) to the geometry of bony prominences and the relative thickness of the different soft tissue elements. Of particular interest, the sensitivity analysis showed that the maximum shear strain in the muscle tissue was very sensitive (29%) to the radius of curvature of the ischium in the plane perpendicular to the shortest radius of curvature. These results are in line with other results in the literature supporting the widely accepted idea that bone geometry will affect internal stresses and strains occurring under bony prominences (see for example [10, 22] for heel ulcers).

B-mode ultrasound (US) imaging represents a promising alternative for assessing geometrical feature-related risk factor bedside or in a clinic with standard medical equipment. US imaging offers the advantages of being portable, non-invasive, with few contraindications and rapid result interpretation. There is therefore a high interest in the community for developing clinical protocols that are suited to reliable parameter assessment [1, 9, 29]. If the measurement of the adipose and muscle tissue thicknesses in the vicinity of the ischium using US has been shown to be both reliable and highly correlated with MRI assessment [1], the measurement of the radius of curvature of the ischium, was, on the contrary, reported to have a poor inter operator reliability [1, 29].

In this perspective, we propose to evaluate, in this exploratory study, the agreement between US-derived and bi-planar X-ray radiography-derived measurements of geometrical features of the Ischial Tuberosity (IT) in a realistic loaded sitting position in healthy volunteers, to quantify the global standard deviation of reproducibility using the ISO standard 2725–2 and to quantify the influence of the angular position of the pelvis in the seated position on the US-based assessment of radii of curvature.

2   Methods

For the sake of clarity, only the experimental material previously acquired in the studies of [24, 25] and pertaining to the current study are briefly recalled hereunder in Sect. 2.1.

2.1 Participants and Protocol

Data of 13 healthy subjects (8 men and 5 women; age: 26 ± 5 years, weight: 70 ± 9 kg, BMI: 22.6 ± 3.4 kg/m2) previously acquired in the studies of [24, 25] were used.

Biplanar radiographs were taken in frontal and sagittal views using the EOS low-dose imaging device (EOS imaging, Paris, France). The US acquisition of the subdermal tissue in the region beneath the IT was performed using a commercial device (Aixplorer, SuperSonic Imagine, France) with a linear US probe of 8 MHz central frequency (SuperLinear SL 15–4). A custom-made stool specifically designed for the experiment allowed to fix the US probe at the level of the seat inside the EOS cabin (Fig. 1a). A cross-shaped notch was made in the seat to fix the US probe in two orthogonal positions.

The subject was asked to sit on the stool in the EOS cabin and, with the help of the on-screen display of the US device, was instructed to self-adjust the position of his/her ischium with the center of the screen. Two sets of data were then acquired with the subject in the loaded sitting position: the first one with the US probe in the frontal plane and the second one with the US probe in the sagittal plane. For each acquisition, a pair of bi-planar X-rays (EOS imaging, Paris, France) were acquired (with a pixel size of 0.179×0.179 mm) (Fig. 1b) immediately followed by a US video clip (pixel size of 0.085×0.085 mm), during which the subject was asked to slowly unload his weight with his arms while visually keeping the ischium as aligned as possible with the probe (Fig. 1b–c). This allowed to measure soft tissue thickness in the unloaded position. An additional pair of radiographs was also acquired in the standardized free standing position [8] for the purpose of 3D reconstruction of the pelvis from the EOS radiographs according to the procedure developed previously by Mitton et al. [26].

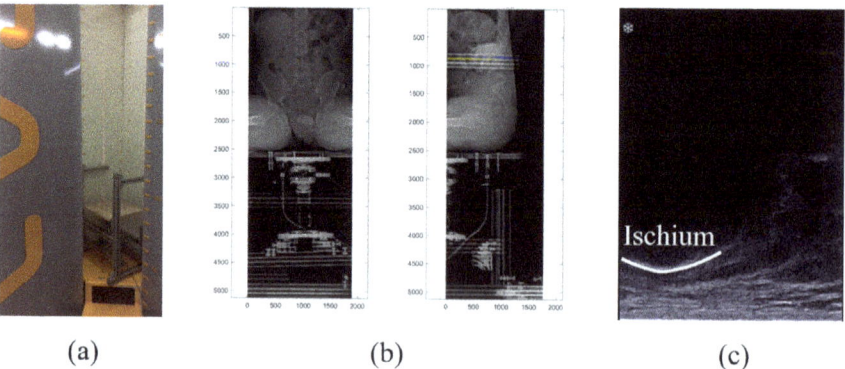

(a)   (b)   (c)

Fig. 1. **a** Custom-made stool allowing to fix the US probe at the level of the seat inside the EOS cabin **b** bi-planar X-ray radiographies acquired in the loaded sitting position **c** US image in sitting loaded configuration with indication of ischium contour

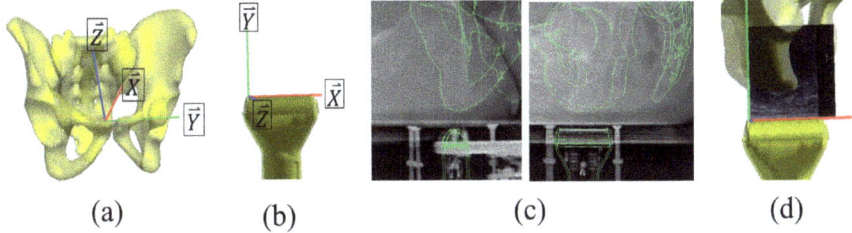

(a)   (b)   (c)   (d)

Fig. 2. a 3D reconstruction of the pelvis and associated coordinate system **b** 3D CAD model of the linear US probe and associated coordinate system **c** cropped frontal and sagittal radiographs in the loaded sitting position with projection of (i) 3D reconstruction of the pelvis and (ii) 3D CAD model of the probe respectively **d** Bimodal image registration

2.2 Data Analysis

2.2.1 Bimodal Image Registration

3D reconstruction of the pelvis (Fig. 2a) was performed from the EOS radiographs in the standing position according to the procedure developed previously by Mitton et al. [26]. The 3D subject-specific model of the pelvis was then projected on the frontal and sagittal radiographs in the loaded sitting positions (in both the acquisition with the US probe in the frontal plane and the acquisition with the US probe in the sagittal plane). The position of the pelvis was manually adjusted until the contours matched those of the radiographs (Fig. 2c). The pelvic coordinate system (Fig. 2a) was defined from the left and right acetabula and S1 endplate of the 3D reconstruction following the definition given by Dubois [7].

Similarly, the 3D CAD model of the linear US probe (Fig. 2b) was projected on the frontal and sagittal radiographs in the loaded sitting positions. The position was manually adjusted until the contours matched those of the radiographs (Fig. 2c). The probe coordinate system (Fig. 2b) was defined from the corners of the top part of the probe (transducer array). The 2D US image was positioned in the 3D EOS cabin space using the probe coordinate system (Fig. 2d).

2.2.2 Assessment of Morphological Parameter

US-based assessment of radii of curvature (US-RoC) and soft tissue thickness (US-STT). In both acquisitions (US probe in the frontal and sagittal planes respectively), 10 points were manually selected on the lowest part of the ischial tuberosity border visible on the US image in the seated position (Fig. 3a). The radius of curvature (Fig. 3b) was computed using a least-squares regression procedure using a custom-made MATLAB subroutine (MathWorks, Natick, MA, USA). The height of the selection, denoted Δz, was defined as the difference between maximum-altitude and the minimum-altitude points (Fig. 3a). The vertical component of the distance

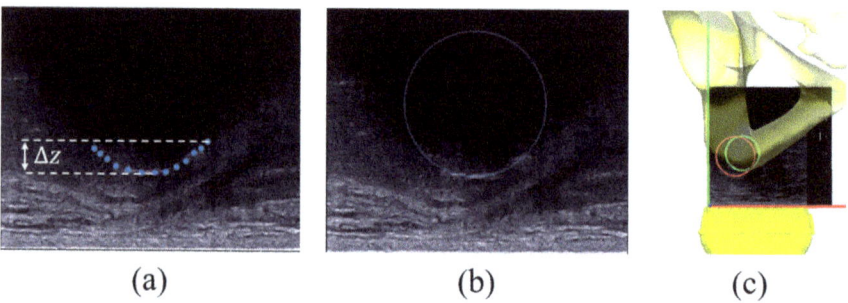

Fig. 3 **a** Points manually selected on the inferior border of each IT **b** least squares fitting circle **c** US-based and EOS-based least squares fitting circles (respectively in red and green) superposed on US image and 3D reconstruction of pelvis in the EOS cabin (global coordinate system)

between lowest point of the IT and overlying skin surface was used to define the soft tissue thickness.

EOS-based assessment of the radii of curvature (EOS-RoC) and soft tissue thickness (EOS-STT). In both acquisitions (US probe in the frontal and sagittal planes respectively), the pelvis surface mesh intersecting points with the US probe plane were computed using a custom-made MATLAB subroutine (MathWorks, Natick, MA, USA). The points in the range Δz (previously determined for each configuration and for each subject) from the minimum-altitude point were selected. The radius of curvature was computed using the same least-squares regression procedure as for the US-based data psoints. Both the US-based and EOS-based least squares fitting circles were superposed on US image and 3D reconstruction of pelvis in the EOS cabin (global) coordinate system (Fig. 3c). The vertical component of the distance between the lowest point of the pelvis surface mesh intersecting with the US probe plane and overlying skin surface was used to define the soft tissue thickness.

2.2.3 Statistical Analysis

On a subset of 4 subjects (26 ± 1 year old), the inter-observer reproducibility standard deviation (SD_r) of the US-based assessment of radii of curvature was computed using the method described in the ISO standard 5725 [14]. Each US image was processed 3 times by 3 operators in both acquisitions (US probe in the frontal and sagittal planes respectively). Reproducibility was estimated by computing the 95% confidence interval (95%CI) ($= 2 \cdot SD_r$).

The agreement between the US-based and EOS-based assessment of radii of curvature of the IT was described graphically with a Bland–Altman plot with mean of differences, reported with corresponding 95% confidence interval (CI), and lower and upper limits of agreement, calculated as mean $\pm 1.96 \times$ standard deviation. Differences were assessed using a Wilcoxon-Signed-Rank Test (paired data) at the default 5% significance level and was further described.

2.2.4 Impact of Pelvis Angular Position on the Computed Radii of Curvature

A sensitivity study was performed to quantify the influence of the angular position of the pelvis in the seated position on the US-based assessment of radii of curvature.

Prior to this step, the maximum difference in the relative angular position of each pelvis to that of the average pelvic pose (averaged position and orientation) was quantified as follows: For each acquisition, the angular position of the pelvis frame was expressed in the EOS cabin (global) coordinate system. The orientation matrix of the relative angular position of each pelvis to that of the average pelvic pose was calculated. The decomposition of the rotation matrix was done using the XYZ rotation sequences of Cardan angles. The averaged position and orientation of all the pelvic frames was computed.

The procedure for the sensitivity study was the following: Each pelvis was repositioned in the averaged pelvic position (i.e. transformed so that the pelvis frame is aligned with that of the average pelvic pose). Then, for each pelvis, the pelvic obliquity (rotation around the global X-axis) and the anterior/posterior tilt (rotation around the global Y-axis) were modified independently by an increment of 5°, 10° and 15°. For each configuration, the (EOS-based) radius of curvature was computed according to the procedure described in Sect. 2.2.2 (Fig. 4).

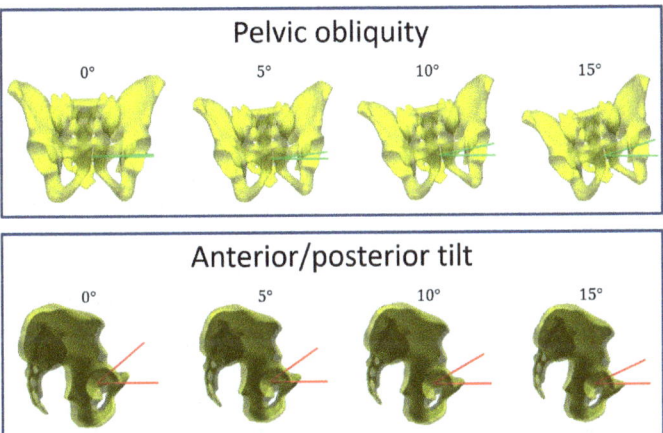

Fig. 4 To assess the impact of small perturbations of the pelvic angular position during seating on the radius of curvature estimated using the procedure described in Sect. 2.2.2, both **a** the pelvic obliquity and **b** the anterior/posterior tilt were modified independently by an increment of 5°, 10° and 15° and the radius of curvature was computed

3   Results

Out of the 13 subjects included, the ischial tuberosity border was visible on the US image of 7 healthy subjects only (4 men and 3 women; age: 28 ± 6 years, weight: 66 ± 7 kg, BMI: 21.6 ± 2.2 kg/m2). The results presented in this section therefore focuses on these 7 subjects only.

3.1   Assessment of Morphological Parameter

The reproducibility of the US-based assessment of radii of curvature was 1.7 mm (95% CI = 3.5 mm) in the frontal plane and 1.3 mm (95% CI = 2.6 mm) in the sagittal plane respectively.

The results of the bimodal image registration and of the assessment of the radii of curvature and soft tissue thickness in both acquisitions (US probe in the frontal and sagittal planes respectively) are given in Table 1 below together with the results of the US-based and EOS-based assessment of Radii Of Curvature (RoC) and Soft Tissue Thickness (STT).

Figure 5 below shows the Bland–Altman plot describing the agreement between US-based and EOS-based assessment of radii of curvature in the frontal and sagittal planes respectively. The bias or mean of differences were $+ 3.3$ mm and $- 5.7$ mm in the frontal and sagittal planes respectively. The mean differences are not zero in either case and this means that, on average, the US measures 3.3 mm more than the EOS-based in the frontal plane (overestimation) and 5.7 less in the sagittal plane (underestimation) respectively. As can be further seen, the differences were of both positive and negative values. The corresponding 95% CI in the frontal and sagittal planes were respectively 1.8 mm and 3.7 mm. The lower and upper limits of agreement were respectively −6.2 mm and 12.8 mm in the frontal plane. The lower and upper limits of agreement were respectively −25 mm and 13.6 mm in the sagittal plane.

Results of the Wilcoxon signed-rank-test, however did not allow rejecting the null hypothesis about the differences between the US-based and EOS-based ROC (p values of 0.16 and 0.30 in the frontal and sagittal planes respectively). Determining soft tissue thickness in the unloaded position was not always possible because the IT was not systematically visible on the US images.

3.2   Impact of Pelvis Angular Position on the Computed Radii of Curvature

The maximum difference in the relative angular position of each pelvis to that of the average pelvic pose was $17°$ (anterior posterior tilt). This allows to estimate an

Table 1 Results of the bimodal image registration and of the assessment of the radii of curvature and soft tissue thickness in both acquisitions. US-based (red) and EOS-based (green) least squares fitting circles are superposed on US image and 3D reconstruction of pelvis in the EOS cabin (global coordinate system). All dimensions are in mm

	Probe in the frontal plane (mm)			Probe in the sagittal plane (mm)		
1		US-ROC	12.2		US-ROC	12.5
		EOS-RoC	15.4		EOS-RoC	23.9
		Difference	−3.2		Difference	−11.4
		US-STT	11.7		US-STT	12.2
		EOS-STT	14.4		EOS-STT	14.5
		Difference	−2.7		Difference	−2.3
2		US-ROC	13.9		US-ROC	18.8
		EOS-RoC	12.3		EOS-RoC	13.6
		Difference	1.6		Difference	5.2
		US-STT	11.6		US-STT	16.1
		EOS-STT	11.5		EOS-STT	16.3
		Difference	0.1		Difference	−0.2:
3		US-ROC	16.5		US-ROC	13.1
		EOS-RoC	7.4		EOS-RoC	11.7
		Difference	9.1		Difference	1.4
		US-STT	12.8		US-STT	12.6
		EOS-STT	14		EOS-STT	14.8
		Difference	−1.2		Difference	−2.2
4		US-ROC	20.4		US-ROC	8.5
		EOS-RoC	8.6		EOS-RoC	6.0
		Difference	11.8		Difference	− 2.7
		US-STT	16.3		US-STT	10.2
		EOS-STT	12.6		EOS-STT	21.5
		Difference	3.7		Difference	− 11.3
5		US-ROC	8.6		US-ROC	7.2
		EOS-RoC	6.8		EOS-RoC	9.9
		Difference	1.8		Difference	−2.7
		US-STT	15.7		US-STT	13.9
		EOS-STT	16.6		EOS-STT	19.1
		Difference	−0.9		Difference	−5.2
6		US-ROC	9		US-ROC	11.4
		EOS-RoC	8.4		EOS-RoC	36.8
		Difference	0.6		Difference	−25.4
		US-STT	13.3		US-STT	16.4

(continued)

Table 1 (continued)

		EOS-STT	16.3		EOS-STT	18.1
		Difference	− 3		Difference	−1.7
7		US-ROC	14		US-ROC	6.9
		EOS-RoC	12.7		EOS-RoC	16.2
		Difference	1.3		Difference	−9.3
		US-STT	8.8		US-STT	8.7
		EOS-STT	11.7		EOS-STT	11.2
		Difference	−2.9		Difference	−2.5

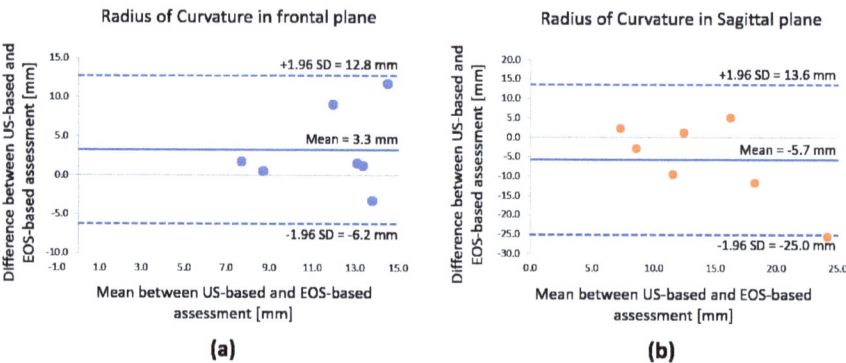

Fig. 5 Plot of differences between US-based and EOS-based assessment of radii of curvature versus the mean of the two measurements (data from Table 1) in the frontal **a** and sagittal **b** planes respectively

order of magnitude of the inter-individual variability in pelvis pose in realistic loaded sitting position in healthy volunteers.

Increasing the pelvic obliquity, lead, on average, to a decrease in the EOS-based radius of curvature assessed (0.80, 1.5 and 18 mm respectively for a rotation of 5°, 10° and 15° in the frontal plane; 0.2, 0.6 and 0.1 mm respectively in the sagittal plane for a rotation of 5°, 10° and 15°).

Increasing the anterior/posterior tilt, on average lead to a relatively small increase in the EOS-based radius of curvature assessed (0.80, 1.4 and 2.5 mm respectively for a rotation of 5°, 10° and 15° respectively in the frontal plane; 1.4, 1.7 and 2.7 mm respectively in the sagittal plane for the 3 rotations).

4   Discussion

The objective of this exploratory study was to evaluate the agreement between Ultrasound-based and bi-planar X-Ray radiography-based assessment of the geometrical features of the Ischial Tuberosity in the context of the prevention of seating-related pressure injury. Assessment of the geometry of bony prominences is important for PU prevention because the local mechanical environment within soft tissues—which has been shown to be a correlated to soft tissue injury risk [4, 20]—is very sensitive (in the statistical sense i.e. variance-based sensitivity analysis) to the geometry of bony prominences [10, 22, 24]. Based on these results, establishing routine ultrasound scans of the ischium could potentially lead to development of new ultrasound-based risk assessment tools that are more specific for identifying susceptibility to seating-related pressure injuries, particularly considering the individual's internal anatomy. This probably explains the increasing enthusiasm of the community for the measurement of this geometrical feature-related risk factor using medical imaging.

Results obtained in this contribution on $N = 7$ healthy however suggest that more effort is needed to establish and standardise optimal measurement and processing procedures and test protocols for the assessment of geometrical features of the IT using US. On average, US-based assessment of radius of curvature underestimated that of the EOS-based in the frontal plane and overestimated in the sagittal plane respectively. This may be partly due to the fact that approximating the ischial tuberosity by a torus (two radii of curvature in the frontal and sagittal planes which are not necessarily aligned with the IT orientation) is too gross and leads to biases in the assessment of morphological parameters. This can also be explained by the ultrasound imaging artefacts inherent to this technology, which, make the detection of bone contours less robust. The fitting of the circle is also dependent on the pelvic posture in sitting. If participants are sitting forward or backward, the circumference of the lowest point of the IT changes dramatically and it is very difficult to measure since the IT is flattened.

In the literature, Swaine and colleagues (2008) represented the ischium using two radii of curvature (along anatomical planes, representing the shortest and longest axis)—other authors limiting the analysis to only one plane [1]. If a good inter-rater reliability for soft tissue thickness assessment of the soft tissue layers overlying inferior curve of IT (skin & fat, tendon & muscle, while thickness) are generally reported [29], observed a poor inter-rater reliability of sonographers in measuring diameter of the inferior curve of IT. This result is in contrast to that reported by Akins et al. [1] who reported an higher concordance value (ICC = 0.712). However, this earlier study utilised a single scan operator to acquire images and two operators to post-process the data on the same images. Such a post processing protocol will inevitably impact the decisions an operator must make during real-time image acquisition and measurement. In this study, a custom-made stool was designed to eliminate the uncertainty related to exclude the impact of inter-operator variability in transducer rotation, medial/lateral and cranial/caudal tiling and pressure exerted over the tissues during examinations. Yet, the reproducibility of the US-based assessment of radii of curvature were relatively high (95% CI = 3.5 mm in the frontal plane and 95% CI = 2.6 mm in the sagittal plane respectively), highlighting the effect of the procedure for extracting morphologic parameters.

The results obtained in our study for US-based assessment of radius of curvature are 13.5 ± 4.1 mm (range [8.6–20.4 mm]) in the frontal plane and 11.2 ± 4.2 mm (range [6.9–18.8 mm]) in sagittal plane respectively. This is within the same range as reported by Akins et al. [1] (between 8.5 mm and 20.0 mm) and [29] (24.2 ± 5 mm in the short axis, 30.0 ± 8 mm in the long axis, respectively. Out of the 13 subjects included in the studies of [24, 25], 6 were excluded for the assessment of the IT curvature in this study since the ischium was not visible due to thick tissue layer of subjects. This can emphasize US-based measurement of IT curvature is highly dependent not only on operator but also the subject. The use of a linear probe represents a limitation of the current study since it is not adapted it has limited penetration. As a perspective work, the use of a curvilinear ultrasound probe will be explored to allow for imaging patients where the image depth penetration and width capture requirements are greater, such as young trauma patients with high muscle bulk, morbidly obese patients with significant subcutaneous fat and patients with fluid overload.

The results obtained in this study for EOS-based assessment of radius of curvature are 10.2 ± 3.2 mm (range [6.8–15.4 mm]) in the frontal plane and 16.9 ± 10.4 mm (range [6.0–36.8 mm]) in sagittal plane respectively. A plausible explanation for the discrepancy between US-based and EOS-based assessment of radii of curvature is the level of accuracy if 3D-reconstructions from bi-planar X-ray images comparaed to CT-scan data. It has been reported that shape differences between 3D models obtained from bi-planar X-rays and CT-scan are, on average 1.6 mm [26].

Recent studies suggest that B-mode Ultrasound imaging constitutes a promising alternative that could overcome some limitations of MRI. Further investigations need to be done in order to estimate the system's overall accuracy in a controlled laboratory setting using precisely built phantom. To make a conclusion on the potential clinical accuracy, the differences between the clinical and laboratory settings must be carefully examined.

References

1. Akins, J. S., Vallely, J. J., Karg, P. E., Kopplin, K., Gefen, A., Poojary-Mazzotta, P., & Brienza, D. M. (2016). Feasibility of freehand ultrasound to measure anatomical features associated with deep tissue injury risk. *Medical Engineering & Physics, 38*, 839–844. https://doi.org/10.1016/j.medengphy.2016.04.026.
2. Al-Dirini, R. M. A., Reed, M. P., Hu, J., & Thewlis, D. (2016). Development and validation of a high anatomical fidelity FE model for the buttock and thigh of a seated individual. *Annals of Biomedical Engineering, 44*, 2805–2816. https://doi.org/10.1007/s10439-016-1560-3.
3. Bouten, C. V., Oomens, C. W., Baaijens, F. P., & Bader, D. L. (2003). The etiology of pressure ulcers: Skin deep or muscle bound? *Archives of Physical Medicine and Rehabilitation, 84*, 616–619. https://doi.org/10.1053/apmr.2003.50038.

4. Ceelen, K. K., Stekelenburg, A., Loerakker, S., Strijkers, G. J., Bader, D. L., Nicolay, K., Baaijens, F. P. T., & Oomens, C. W. J. (2008). Compression-induced damage and internal tissue strains are related. *Journal of Biomechanics, 41*, 3399–3404. https://doi.org/10.1016/j.jbiomech.2008.09.016.

5. Coleman, S., Gorecki, C., Nelson, E. A., Closs, S. J., Defloor, T., Halfens, R., Farrin, A., Brown, J., Schoonhoven, L., & Nixon, J. (2013). Patient risk factors for pressure ulcer development: Systematic review. *International Journal of Nursing Studies, 50*, 974–1003. https://doi.org/10.1016/j.ijnurstu.2012.11.019.

6. Demarré, L., Van Lancker, A., Van Hecke, A., Verhaeghe, S., Grypdonck, M., Lemey, J., Annemans, L., & Beeckman, D. (2015). The cost of prevention and treatment of pressure ulcers: A systematic review. *International Journal of Nursing Studies, 52*, 1754–1774. https://doi.org/10.1016/j.ijnurstu.2015.06.006.

7. Dubois, G. (2014). Contribution à la modélisation musculo-squelettique personnalisée du membre inférieur combinant stéréoradiographie et ultrason. (phdthesis). Ecole nationale supérieure d'arts et métiers—ENSAM.

8. Faro, F. D., Marks, M. C., Pawelek, J., & Newton, P. O. (2004). Evaluation of a functional position for lateral radiograph acquisition in adolescent idiopathic scoliosis. *Spine, 29*, 2284–2289.

9. Gabison, S., Hayes, K., Campbell, K. E., Swaine, J. M., & Craven, B. C. (2019). Ultrasound imaging of tissue overlying the ischial tuberosity: Does patient position matter? *Journal of Tissue Viability*. https://doi.org/10.1016/j.jtv.2019.07.001.

10. Gefen, A. (2010). The biomechanics of heel ulcers. *Journal of Tissue Viability, 19*, 124–131. https://doi.org/10.1016/j.jtv.2010.06.003.

11. Gefen, A., van Nierop, B., Bader, D. L., & Oomens, C. W. (2008). Strain-time cell-death threshold for skeletal muscle in a tissue-engineered model system for deep tissue injury. *Journal of Biomechanics, 41*, 2003–2012. https://doi.org/10.1016/j.jbiomech.2008.03.039.

12. Ghostine, B., Sauret, C., Assi, A., Bakouny, Z., Khalil, N., Skalli, W., & Ghanem, I. (2017). Influence of patient axial malpositioning on the trueness and precision of pelvic parameters obtained from 3D reconstructions based on biplanar radiographs. *European Radiology, 27*, 1295–1302. https://doi.org/10.1007/s00330-016-4452-x.

13. Gray, R. J., Voegeli, D., & Bader, D. L. (2016). Features of lymphatic dysfunction in compressed skin tissues—Implications in pressure ulcer aetiology. *Journal of Tissue Viability, 25*, 26–31. https://doi.org/10.1016/j.jtv.2015.12.005.

14. ISO, 1994. (1994). ISO 5725–2 Accuracy (Trueness and Precision) of Measurements Methods and Results.—Part 2: Basic Method for the Determination of Repeatability and Reproducibility of a Standard Measurement Method, International Organisation for Standardisation, Geneva, Switzerland. Retrieved from https://www.iso.org/standard/69419.html.

15. Jiang, L., Tu, Q., Wang, Y., & Zhang, E. (2011). Ischemia-reperfusion injury-induced histological changes affecting early stage pressure ulcer development in a rat model. *Ostomy/Wound Management, 57*, 55–60.

16. Levy, A., Kopplin, K., & Gefen, A. (2013). Simulations of skin and subcutaneous tissue loading in the buttocks while regaining weight-bearing after a push-up in wheelchair users. *Journal of the Mechanical Behavior of Biomedical Materials, 28*, 436–447. https://doi.org/10.1016/j.jmbbm.2013.04.015.

17. Linder-Ganz, E., Shabshin, N., Itzchak, Y., & Gefen, A. (2007). Assessment of mechanical conditions in sub-dermal tissues during sitting: A combined experimental-MRI and finite element approach. *Journal of Biomechanics, 40*, 1443–1454. https://doi.org/10.1016/j.jbiomech.2006.06.020.

18. Linder-Ganz, E., Shabshin, N., Itzchak, Y., Yizhar, Z., Siev-Ner, I., & Gefen, A. (2008). Strains and stresses in sub-dermal tissues of the buttocks are greater in paraplegics than in healthy during sitting. *Journal of Biomechanics, 41*, 567–580. https://doi.org/10.1016/j.jbiomech.2007.10.011.

19. Linder-Ganz, E., Yarnitzky, G., Yizhar, Z., Siev-Ner, I., & Gefen, A. (2009). Real-time finite element monitoring of sub-dermal tissue stresses in individuals with spinal cord injury: Toward prevention of pressure ulcers. *Annals of Biomedical Engineering, 37*, 387–400. https://doi.org/10.1007/s10439-008-9607-8.
20. Loerakker, S., Manders, E., Strijkers, G. J., Nicolay, K., Baaijens, F. P. T., Bader, D. L., & Oomens, C. W. J. (2011). The effects of deformation, ischemia, and reperfusion on the development of muscle damage during prolonged loading. *Journal of Applied Physiology, 111*, 1168–1177. https://doi.org/10.1152/japplphysiol.00389.2011.
21. Luboz, V., Bailet, M., Boichon Grivot, C., Rochette, M., Diot, B., Bucki, M., & Payan, Y. (2018). Personalized modeling for real-time pressure ulcer prevention in sitting posture. *Journal of Tissue Viability, 27*, 54–58. https://doi.org/10.1016/j.jtv.2017.06.002.
22. Luboz, V., Perrier, A., Bucki, M., Diot, B., Cannard, F., Vuillerme, N., & Payan, Y. (2015). Influence of the calcaneus shape on the risk of posterior heel ulcer using 3D patient-specific biomechanical modeling. *Annals of Biomedical Engineering, 43*, 325–335. https://doi.org/10.1007/s10439-014-1182-6.
23. Luboz, V., Petrizelli, M., Bucki, M., Diot, B., Vuillerme, N., & Payan, Y. (2014). Biomechanical modeling to prevent ischial pressure ulcers. *Journal of Biomechanics, 47*, 2231–2236. https://doi.org/10.1016/j.jbiomech.2014.05.004.
24. Macron, A., Pillet, H., Doridam, J., Rivals, I., Sadeghinia, M. J., Verney, A., & Rohan, P.-Y. (2020). Is a simplified Finite Element model of the gluteus region able to capture the mechanical response of the internal soft tissues under compression? *Clinical Biomechanics, 71*, 92–100. https://doi.org/10.1016/j.clinbiomech.2019.10.005.
25. Macron, A., Pillet, H., Doridam, J., Verney, A., & Rohan, P.-Y. (2018). Development and evaluation of a new methodology for the fast generation of patient-specific Finite Element models of the buttock for sitting-acquired deep tissue injury prevention. *Journal of Biomechanics, 79*, 173–180. https://doi.org/10.1016/j.jbiomech.2018.08.001.
26. Mitton, D., Deschênes, S., Laporte, S., Godbout, B., Bertrand, S., de Guise, J. A., & Skalli, W. (2006). 3D reconstruction of the pelvis from bi-planar radiography. *Computer Methods in Biomechanics and Biomedical Engineering, 9*, 1–5. https://doi.org/10.1080/10255840500521786.
27. Moerman, K. M., van Vijven, M., Solis, L. R., van Haaften, E. E., Loenen, A. C. Y., Mushahwar, V. K., & Oomens, C. W. J. (2017). On the importance of 3D, geometrically accurate, and subject-specific finite element analysis for evaluation of in-vivo soft tissue loads. *Computer Methods in Biomechanics and Biomedical Engineering, 20*, 483–491. https://doi.org/10.1080/10255842.2016.1250259
28. Oomens, C. W. J., Bader, D. L., Loerakker, S., & Baaijens, F. (2015). Pressure induced deep tissue injury explained. *Annals of Biomedical Engineering, 43*, 297–305. https://doi.org/10.1007/s10439-014-1202-6.
29. Swaine, J. M., Moe, A., Breidahl, W., Bader, D. L., Oomens, C. W. J., Lester, L., O'Loughlin, E., Santamaria, N., & Stacey, M. C. (2018). Adaptation of a MR imaging protocol into a real-time clinical biometric ultrasound protocol for persons with spinal cord injury at risk for deep tissue injury: A reliability study. *J. Tissue Viability, Seating Biomechanics, 27*, 32–41. https://doi.org/10.1016/j.jtv.2017.07.004.
30. Zeevi, T., Levy, A., Brauner, N., & Gefen, A. (2017). Effects of ambient conditions on the risk of pressure injuries in bedridden patients-multi-physics modelling of microclimate. International Wound Journal. https://doi.org/10.1111/iwj.12877.

Characterising the Soft Tissue Mechanical Properties of the Lower Limb of a Below-Knee Amputee: A Review

Seyed Sajad Mirjavadi, Andrew J. Taberner, Martyn P. Nash, and Poul M. F. Nielsen

Abstract This paper presents a review of existing biomechanical models and technologies relevant to the study of soft tissues of below-knee amputees. Special attention is paid to related studies on biomechanical measurements of soft tissues, including the approaches for measuring forces and pressure distributions between stump and socket, and also estimating the deformations of soft tissues under loads. Mechanical properties of soft tissues related to their time-dependent and large deformation behaviours are discussed in the context of viscoelasticity and hyperelasticity. We review techniques used to measure the shapes of residual limbs, including volume fluctuations. We also survey methods for measuring tissue deformation using magnetic resonance imaging (MRI) and camera-based devices. The motivation for this review is to highlight techniques and methods for characterising soft tissues of below-knee amputees, and to discuss their limitations in order to guide future studies.

Keywords Soft tissue · Below-knee amputation · Lower-limb stump · Biomechanical properties · Tissue deformation

1 Introduction

There are two main approaches to characterising the geometry and mechanics of an amputee's stump. One approach relies on the knowledge of the outer geometry of the stump, and how it deforms under various loading conditions. By applying a range of forces and torques to the skin, the resulting surface displacement fields can be measured and used to infer the surface mechanical properties of the stump. An alternative approach is to fully characterise the anatomy and biomechanical properties of the constituent tissues, such as the skin, muscles, ligaments, tendons, and bones that comprise the stump [1].

S. S. Mirjavadi (✉) · A. J. Taberner · M. P. Nash · P. M. F. Nielsen
Auckland Bioengineering Institute, University of Auckland, Auckland, New Zealand
e-mail: seyedsajadmirjavadi@gmail.com

A. J. Taberner · M. P. Nash · P. M. F. Nielsen
Department of Engineering Science, University of Auckland, Auckland, New Zealand

© The Author(s), under exclusive license to Springer Nature Switzerland AG 2021
K. Miller et al. (eds.), *Computational Biomechanics for Medicine*,
https://doi.org/10.1007/978-3-030-70123-9_8

In this paper, we review techniques used by researchers to characterise the various tissue components using imaging techniques, such as MRI. We assess some of the techniques for modelling deformations under various loading conditions, and the use of these data for characterising the mechanical properties of the stump.

2 Anatomy and Biomechanical Measurements of the Stump of Below-Knee Amputees

The first step in modelling biomechanics of the stump is to characterise the geometry and constitutive properties of its tissue components. The bony structure of the stump includes the femur, tibia, fibula, and patella or kneecap. Interactions of these bones through the knee joint are mediated by cartilage, medial/lateral menisci, ligaments/tendons (including the fibular/medial collaterals, anterior/posterior cruciates, and patellar tendon), muscles (such as deep muscle, peroneus brevis muscle, hamstrings and quadriceps), skin, adipose tissues, and synovial fluid.

Biomechanical characterisation of soft tissues requires techniques to apply and measure loads, such as distributions of pressures, shear forces, and torques. These analyses also require estimates of the unloaded geometries, measurements of the deformations of the soft tissues under loads, and estimates of the remote loading conditions and kinematic constraints surrounding the stump.

2.1 Characterisation of the Reference Geometry and Deformations

2.1.1 Magnetic Resonance Imaging (MRI)

MRI has been applied in radiology as a medical imaging technique to acquire images of the anatomy and physiological processes of the human body. An advantage of MRI is the absence of X-rays and ionising radiation. Although MRI provides useful information about the geometry of soft tissue, the scanners are costly to purchase and maintain, and typically require greater technical skill to operate compared to most other imaging techniques [2]. Moreover, due to the use of strong magnetic fields, MRI cannot be performed on patients with implanted medical devices, such as neuro-stimulators, pacemakers, or drug infusion pumps.

To characterise of the geometry of a lower-limb stump fitted with a prosthetic socket, Cagle et al. [3] replaced the MRI non-compatible parts of the socket with polymeric fibre and epoxy-acrylic resin. They reported that the polymer composite was not detectable in the MR images. To improve image quality, the researchers used ultrasound gel to fill the voids such as the distal gap between the socket and prosthetic liner.

2.1.2 Camera-Based Digital Image Correlation (DIC)

Accurate measurements of deformations are essential to investigate and understand the mechanical response of tissues subjected to loads. Furthermore, effective prosthetic socket designs for below-knee amputees rely on the accurate characterisation of the geometries of the residual limbs, their shape, and volume fluctuations [4]. To achieve this, camera-based digital image correlation (DIC) techniques [5, 6] and Digital Volume Correlation (DVC) [7, 8] are frequently applied for in vivo measurement of deformations of soft tissues such as skin.

Three-dimensional DIC (3D-DIC) is an optical-numerical technique to evaluate the dynamical mechanical behaviour (deformation over time) at the surfaces of an object, such as soft tissue. This technique can be applied for high resolution extraction of the shapes and displacements at different length scales. A range of commercial and academic 3D-DIC software tools have been developed. For multi-view analyses, which are especially favourable in biomedical applications, it is helpful to use 3D-DIC packages which offer straightforward calibration and data-merging capabilities.

Solav et al. [9] presented the MultiDIC technique, suitable for multi-view setups. For reconstructing the dynamical behaviour of surfaces from multiple stereo-camera pairs, MultiDIC integrates robust 2D-DIC software with effective calibration procedures. They have developed a 12-camera setup to test the efficiency of the MultiDIC method. One limitation of this study was the use of cheap camera modules to test the efficiency of the method. Accordingly, the precision reported was modest. Future investigations should characterise the performance of MultiDIC using higher quality optics. Use of only a single multi-view setup and calibration object, as well as the fact that the technique does not provide a real-time displacement or strain measurements, also limits the applications of this approach.

Using an image-based system, Solav et al. [4] proposed a new approach to measure residual limb deformation with a multi-camera system to acquire simultaneous images of the stump surface. To acquire accurate time-varying deformation fields of the stump surface, the analyses of these images were conducted using an open-source 3D-DIC toolbox. The researchers obtained the measurements on transtibial amputee stumps during knee flexion, swelling upon socket removal, and muscle contraction. There were four main limitations of this study. Firstly, the approach is a time-intensive process, requiring that the limb be speckled with ink before the measurements are made. Secondly, the measurement could not be performed when the stump was inside the socket, since the method is optical in nature. Thirdly, the synchronisation latency among the cameras ($\sim$30 ms), contributing to the 3D reconstruction error, had not been accounted for. This could be addressed by substituting the software-controlled synchronisation with hardware triggers. Fourthly, results were obtained for only a single subject, so further research is required for verification of the results reported in the study.

2.1.3   Sensors for Measuring Single Point Displacements for Static or Quasi-Static Samples

Flynn et al. [10] presented an experiment for determining the mechanical charac-teristics of in vivo human skin using a custom-built force-sensitive 3D micro-robot consisting of a probe that could move quickly within a work volume. The micro-robot exerted load on the anterior forearm and the posterior upper arm, and would also be suitable for use on the stump. In another study, Flynn et al. performed an experi-ment in which the facial skin of five individuals was exposed to a set of deformations using the micro-robotic device. The Ogden hyperelastic constitutive relation [11] and quasi-linear viscoelasticity were used for developing a finite element model and simu-lating the experiments. They analysed force-displacement curves of the probe-tip, and concluded that the skin has significantly anisotropic, viscoelastic, and nonlinear mechanical properties. They showed the dependency of the response of the upper arm skin on the orientation of the arm. The primary limitation of these studies was that only normal load was applied to measure the displacement, with non-normal loadings not considered. The researchers also did not consider skin deformations using multiple probes, which limited the types of deformation that could be induced.

Sengeh et al. [12] presented a 3D viscoelastic characterisation of a transtibial stump based on in vivo indentations and MRI data. They used a robotic in vivo inden-tation system for inducing displacements at 18 selected locations. They also evalu-ated the nonlinear elastic/viscoelastic mechanical behaviours of the soft tissues using the inverse finite element method. The nonlinear elastic behaviours were modelled using the Ogden hyperelastic relations for which the strain energy function, W, is presented in Eq. (1), and viscoelastic behaviours were captured according to quasi-linear viscoelasticity. They reported the material parameters for the soft tissues of the stump. However, the indenter used in this research had a geometry with sharp edges, which resulted in convergence problems in simulations. Because of the indenter shape and size, accurate capture of data was difficult in regions of high curvature. Use of a spherical and smaller indenter may help for loading around uneven surfaces. A further limitation of the evaluation represented in this paper is that the tissue defor-mation had not been validated. It may be better to apply the techniques of surface deformation measurement based on camera-based DIC.

$$W = \sum_{p=1}^{N} \frac{\mu_p}{\alpha_p} (\lambda_1^{\alpha_p} + \lambda_2^{\alpha_p} + \lambda_3^{\alpha_p}) \tag{1}$$

where p, μ_p and α_p are material constants and $\lambda_j (j = 1, 2, 3)$ are principal stretches.

2.2 Techniques and Sensors for Measuring the Loads Acting on the Stump

Devices designed for measuring the loads applied to the stump may be categorised as single-point or arrays of sensors. Any force probing will affect neighbouring probes because soft tissues deform significantly in response to the forces that we wish to use, while this issue might not be raised in a single load sensor. In the category of an array of sensors, each sensor might have interference with the other sensor in the array. However, the main reason of using an array of sensors instead of a single sensor is to simultaneously obtain the data from stump-socket interface, while the stump is subjected to the load, and this cannot be implemented by just using a single sensor.

2.2.1 Sensors for Measuring Forces and/or Torques at a Single Point

The Fitsocket Robot devised by a group at the Massachusetts Institute of Technology (MIT) is comprised of a transducer for applying a normal force on the stump, while other transducers are used to constrain the motion of the posterior part of the stump [12]. Flynn et al. [10] used a custom-made microrobot for applying the loads to investigate the effects of shear forces on the skin and obtaining shear stiffness.

2.2.2 Techniques to Measure Spatial Distributions of Loads Acting on the Stump

The stresses distributed at the interface between the prosthesis and residual limb are generally combinations of shear and normal stresses. High stresses at the interface may lead to decreased blood flows and skin-related problems. Thus, measuring shear stress is as significant as the measurement of normal stress. The first measurements of the shear stresses at the stump/socket interface were reported by Appoldt et al. [13], who used tangential pressure transducers embedded in the walls of the socket. This system was limited due to its inability to measure both shear and normal stresses simultaneously.

Sanders and Daly [14] designed a system for simultaneously measuring the stump–socket interface stresses in three orthogonal directions. They positioned the force sensors at different locations inside the prosthetic socket of a below-knee amputee to measure the stresses during walking. The transducers were oriented in orthogonal directions over 6.35 mm diameter sensing areas. The shear stresses at the interface were measured from the estimated differences in bending moments between the strain gauge locations. The normal stresses were measured using full-bridge diaphragm strain gauge networks. Despite being able to measure the shear and normal stresses simultaneously, their use of piston-based transducers for stress measurements was limited by their bulky size and complex instrumentation. Moreover, a source of error arises from the deflection of these transducers under load, leading to the deflection of skin from its initial position, which results in errors in estimating shear stress from their measurements.

Force-sensing resistors (FSRs) can be used for measuring the dynamic stump–socket interface pressure of walking transtibial amputees. Generally, an FSR is a force sensor whose resistance reduces by exerting a normal force. FSRs can be constructed with different shapes and can estimate the changes in applied loads. Stresses are measured by dividing the estimated load by the surface areas of the sensors. Convery and Buis [15] arranged 350 pressure sensors over the inner surface of the socket to estimate the dynamic interface pressure during the gait cycle. They claimed that these compound sensors indicated that the use of their hydro cast prosthetic socket led to lower pressure gradients during dynamic loads compared to a patellar-tendon-bearing (PTB) socket. One limitation of their study was that Convery and Buis [15] investigated the pressure field distribution only during the stance phase of the gait cycle, whereas the swing phase was not studied.

Piezo-resistive FSRs are good candidates for medical applications [16] because they are thin, flexible, and conformable devices. Since they can be configured as thin and flexible sheets, piezo-resistive sensors can be placed within the socket for monitoring the pressures [17]. Arrays of FSRs can be used for monitoring the spatial distributions of stresses at the stump–socket interface. As an example, Ruda et al. [18] arranged five sensors embedded within a thin acetate sheet to monitor the pressures at the stump–socket interface. However, because of the very small surface areas of FSRs, the stress estimates were somewhat unreliable.

Capacitive sensors have also been used for pressure monitoring at the stump–socket interface. A prototype capacitance pressure sensor was introduced by Polliack et al. [19] for prosthetic socket application, where 16 sensors were arranged on a silicone substrate. They reported a mean hysteresis error of $12.93\% \pm 4.63\%$. However, these capacitance pressure sensors were unidirectional and could only be used to estimate the applied pressure. Laszczak et al. [20] developed a novel capacitive sensor for measuring the shear and normal stresses at the stump–socket interface. The design of the sensor was based on the change of the area and height of the sensor's frame, which indirectly affects capacitance within the sensor. The area of applied load on the sensor was 20 mm by 20 mm, and they assumed that the area was under uniform shear and normal stresses, which is not strictly correct in any location at the stump-socket interaction. In particular, parts of the stump or socket with high curvature, will be subject to large errors in the estimated stresses using this sensor. In this study, the application of the sensor was limited to quasi-static behaviour, although there is an obvious demand to develop it to measure deformation over time.

Fibre Bragg grating (FBG) sensors have high sensitivity, durability, immunity to electromagnetic interference, resistance to aggressive environments [21], and have been applied to measure various quantities including pressure, force, strain, temperature, and humidity. The stiffness of the polymer matrix affects the performance of an FBG sensor for stress estimation. Various matrix materials were studied by Al-Fakih et al. [22] to optimise the accuracy of stress measurement at the stump–socket inter-face. The results demonstrated that the use of a harder and thicker matrix material in the socket led to greater accuracy and sensitivity. In another study by the same group [23], FBG elements were placed in thin layers of epoxy-based sensing pads for in-socket stress measurement. The FBG-instrumented epoxy pads were embedded in silicone polymers to make pressure sensors. The efficacy of the FBG-epoxy sensors was tested by inserting and inflating heavy-duty balloons into the sockets using compressed air to simulate the conditions of a transtibial amputee's patellar tendon bar. However, like most piezo-resistive pressure sensors, these FBG-based pressure sensors could only be used to estimate the normal stresses.

3 Identifying the Mechanical Properties of the Tissues of the Lower Limb

Finite element (FE) analysis is an approach widely used in bioengineering to estimate the stresses and strains in complicated mechanical systems, and has proven to be a useful tool for prosthetic socket design. Since the first development of FE models of the transtibial residual limb and prosthetic socket [24], several models have been developed to improve prosthesis design [25]. The stump–prosthesis interface was the focus of the first substantial efforts using computational biomechanics for prosthetic research. The earliest studies used two-dimensional (2D) models to explore the load transfer between the stump and the prosthetic socket [26, 27]. Later studies used 3D models for the stump–prosthesis interface [28–37].

Interface modelling is useful for estimating the stump–prosthesis pressures and stresses, which are important since they relate to potentially measurable quantities, assisting model validation. The interface pressures and stresses in the earliest studies were predicted by applying static loads to the stump [27–31, 33–35, 38, 39]. In some studies, quasi-dynamic analyses have been performed to investigate interface pressures [34]. However, it should be stated that the finite element models developed in cited studies are not fully dynamic models, whereas the interaction between the prosthetic socket and residual limb is a highly dynamic process [40].

The selection of the contact model can have a significant effect on the stump–prosthesis interface simulation. Some studies have addressed the issues related to large relative tangential displacements by using contact elements that differ in their contact partner identification techniques [29], and by application of explicit finite element formulation [34]. In these studies, the coefficient of friction (COF) value was considered in the range of 0.415–0.7 [28–31, 34]. However, other studies have assumed frictionless contact between the stump and the socket [41–43]. Most studies refer to two experimental COF reports: Sanders et al. [41] carried out static COF measurement between skin, socks, and conventional sockets, and Zhang and Mak [42] carried out similar dynamic COF measurements. A limitation of these studies was that

the COF measurements were performed in dry conditions and at ambient tempera-
tures. Future measurements should preferably incorporate the influences of elevated
temperature and the presence of moisture due to sweat at the skin surface, which
have been reported to have significantly variable influences on COF, as explained by
Derler and Gerhardt [43].

Many early investigations [27–31, 37] used linear constitutive relations for
modelling the mechanical response of the soft tissues of the stump. Linear elasticity
is adequate for a material that deforms and recovers in a linear manner in response
to applied loads. For infinitesimal strains, this behaviour can be specified based on
generalised Hooke's law, including the Young's modulus (E) and the Poisson's ratio
($-1 \leq \nu \leq 0.5$). As the ratio of transverse to axial strain, Poisson's ratio is related to
the compressibility of materials. With high water content, soft tissues are generally
assumed to be incompressible ($\nu = 0.5$). However, this limiting value can lead to
numerical instabilities in finite element computations, so researchers have typically
used values $0.45 \leq \nu \leq 0.49$ in linear elasticity models of soft tissues. The mechanical
response of bone has also been represented using linear elasticity, using the values
10 GPa $\leq E \leq$ 15.5 GPa and $0.28 \leq \nu \leq 0.30$ [28, 30, 31, 34, 36, 37, 39], or bone
has been assumed to be rigid in some studies [29, 33]. Many investigations of limb
surface stresses, based on linear elastic models of soft tissues, used the values $E =$
200 kPa and $\nu = 0.49$ following the work of Zhang et al. [44].

Soft tissues generally undergo large strains and exhibit nonlinear stress-strain
behaviour during physiological loading. It is thus important that finite strain theory
should be adopted to describe the large deformation of soft tissues. Owing to the
recent progress in applied material models for the soft tissues, linear elastic simpli-
fications have been broadly replaced by models that include hyperelasticity and
viscoelasticity, albeit subject to the assumptions of homogeneity and isotropy (the
skin exhibits heterogeneous properties, and skin and the muscle, as a composites of
cells and collagen, respond anisotropically to loads). Hyperelasticity postulates the
existence of a strain energy density (SED) function [45], which relates the strains
of tissues to their corresponding stress values [46]. The biomechanical properties of
soft tissues such as skin, fat, and muscle can be well-characterised using hyperelastic
models [47]. Various constitutive relations have been suggested for modelling hyperelastic materials such as Saint Venant–Kirchhoff, neo-Hookean, Mooney-Rivlin,
Ogden, and Yeoh relations [48]. One of the simplest hyperelastic constitutive relations is the Saint Venant–Kirchhoff model, which is an extension of the geometrically linear elastic constitutive relation to the geometrically nonlinear regime. Saint
Venant–Kirchhoff and Mooney-Rivlin models can be used for phenomenological
descriptions of observed behaviour, while the neo-Hookean model is a mechanistic
model derived from arguments about underlying structure of the material. The constitutive parameters of these models can be tailored to reproduce particular stress-
strain behaviours using inverse finite element methods to interpret biomechanical
measurements. Several recent investigations used hyperelasticity to represent and
study soft tissue stresses [33, 34, 36, 38, 49]. Portnoy et al. proposed an approach
that has been adopted by the majority of subsequent investigations [50]. They used

a compressible neo-Hookean model for fat and muscle with various levels of stiffness, an incompressible neo-Hookean model for scar tissue, and an incompressible extended Mooney-Rivlin model for the skin. It should be noted that the identified studies used SED functions taken from general material characterisation studies, rather than patient-specific information.

In general, the predicted maximum stump–socket interface stresses under stance and gait loading were remarkably higher in the earlier studies that assumed linear elastic response of the soft tissues, compared to the stresses predicted by the more recent hyperelastic models. Specifically, the studies based on linear elasticity reported the maximum interface pressures in the range of 90–783 kPa [28, 39], while the studies using hyperelasticity reported the maximum interface pressures in the range of 1.54–119 kPa [34–36]. Most studies applied a normal uniaxial load to reproduce the stance phase and neglected the shear stress, but one source of difference in these pressures may be the various magnitudes of the loads applied to the stump. It should be noted that the earlier studies neglected the effect of pre-stresses applied to the stump. Lee et al. [31] presented a new practical technique for modelling the contact interfaces by considering the friction/slip conditions and pre-stress exerted on the limb within rectified sockets. The pre-stresses were predicted by moving the penetrated limb surface onto the inner surface of the socket. Then, by keeping the pre-stresses, the loading during the stance phase was simulated when the loads were applied on the knee joint. They reported that maximum normal and shear forces decrease over the regions where socket undercuts are created.

Time-dependence of soft tissue loading in stump FE models has received relatively little attention in the research literature. Some investigations have considered the viscoelastic effects related to stress relaxation. Non-viscoelastic models are not able to determine the absorbed/dissipated energy by the prosthesis due to the fact that energy absorbing damping effects have been neglected. Characterisation of the viscoelastic response enables the examination of transient influences of stress relaxation. For example, Portnoy et al. [33] estimated the risks of soft tissue injuries using a Prony series expansion to represent stress relaxation and creep of the soft tissues.

Silver-Thorn [51] conducted in vivo rate-controlled indentation experiments and stress relaxation tests on the soft tissues of the stump. They showed that the bulk soft tissue responses to compressive loads are nonlinear and rate-dependent. Notable stiffness (Young's modulus) variation between five individuals was also observed. They demonstrated that biomechanical simulations of bulk soft tissues require consideration of nonlinear viscoelastic constitutive relations. Tönük and Silver-Thorn [52] evaluated a suitable set of nonlinear viscoelastic material parameters for the soft tissues using a Prony series expansion and a linear Kelvin–Voigt model of viscoelasticity. They identified a range of viscoelastic parameters based on data from creep and relaxation experiments. Sengeh et al. [12] presented a 3D viscoelastic characterisation of a transtibial stump based on in vivo indentations. They evaluated the nonlinear viscoelastic mechanical behaviours of the soft tissues using inverse finite element methods. The nonlinear elastic behaviours were modelled using Ogden hyperelastic relations, and viscoelastic behaviours were captured according to a two-parameter relaxation function. They identified the material parameters for soft tissues of the stump and found less spatial variability in calculated properties for their test participant than Tönük and Silver-Thorn [52] had observed. Further investigations are needed to study the effects of suddenly applied or shock loads, for which viscous damping plays an important role in the transient mechanical behaviour of the stump.

4   Conclusion

This paper has reviewed recent advances in lower limb amputee modelling based on finite element studies that have considered the interface pressures between the residual limbs and prosthetic sockets. Most studies have considered only the effects of normal uniaxial loads, while neglecting shear loads. The majority of studies have been based on static or quasi-static analyses, whereas the interaction between the prosthetic sockets and residual limbs is typically a highly dynamic process. Although there are recent advances in methods of determining biomechanical properties of soft tissues, there is still an urgent need to develop instrumentation and computational tools to enable characterisation of the individual-specific geometry and tissue mechanical properties in lower limb amputees. For the soft tissues of the lower limb, the viscoelastic effects related to time-dependent material behaviour, and the nonlinear, anisotropic hyperelastic mechanical response to the large deformations that occur during physiological loading, should also be considered to develop more realistic predictions of the biomechanical interactions between the stump and socket that can be used to improve the design of prostheses on a per-patient basis.

References

1. Dickinson, A. S., Steer, J. W., & Worsley, P. R. (2017). Finite element analysis of the amputated lower limb: a systematic review and recommendations. *Medical Engineering & Physics, 43,* 1–18.
2. Burnett, S. J., & Stoker, D. J. (1995). Practical limitations of magnetic resonance imaging in orthopaedics. *Curr Orthop, 9*(4), 253–259.
3. Cagle, J. C., Reinhall, P. G., Allyn, K. J., et al. (2018). A finite element model to assess transtibial prosthetic sockets with elastomeric liners. *Medical & Biological Engineering & Computing, 56*(7), 1227–1240.
4. Solav, D., Moerman, K. M., Jaeger, A. M., et al. (2019). A framework for measuring the time-varying shape and full-field deformation of residual limbs using 3-D digital image correlation. *IEEE Transactions on Biomedical Engineering, 66*(10), 2740–2752.
5. Chu, T. C., Ranson, W. F., & Sutton, M. A. (1985). Applications of digital-image-correlation techniques to experimental mechanics. *Experimental Mechanics, 25*(3), 232–244.
6. Pan, B., Qian, K., Xie, H., et al. (2009). Two-dimensional digital image correlation for in-plane displacement and strain measurement: a review. *Measurement Science and Technology, 20*(6),
7. Rankin, K., Steer, J., Paton, J., et al. (2020). Developing an Analogue Residual Limb for Comparative DVC Analysis of Transtibial Prosthetic Socket Designs. *Materials, 13*(18), 3955.

8. Paternò, L., Ibrahimi, M., Gruppioni, E., et al. (2018). Sockets for limb prostheses: A review of existing technologies and open challenges. *IEEE Transactions on Biomedical Engineering, 65*(9), 1996–2010.

9. Solav, D., Moerman, K. M., Jaeger, A. M., et al. (2018). MultiDIC: An open-source toolbox for multi-view 3D digital image correlation. *IEEE Access, 6,* 30520–30535.

10. Flynn, C., Taberner, A., & Nielsen, P. (2011). Mechanical characterisation of in vivo human skin using a 3D force-sensitive micro-robot and finite element analysis. *Biomechanics and Modeling in Mechanobiology, 10*(1), 27–38.

11. Flynn, C., Taberner, A., Nielsen, P., et al. (2013). Simulating the three-dimensional deformation of in vivo facial skin. *Journal of the Mechanical Behavior of Biomedical, 28,* 484–494.

12. Sengeh, D. M., Moerman, K. M., Petron, A., et al. (2016). Multi-material 3-D viscoelastic model of a transtibial residuum from in-vivo indentation and MRI data. *Journal of the Mechanical Behavior of Biomedical, 59,* 379–392.

13. Appoldt, F. A., Bennett, L., & Contini, R. (1970). Tangential pressure measurements in above-knee suction sockets. *Bulletin of Prosthetics Research, 10*(13), 70.

14. Sanders, J. E., & Daly, C. H. (1993). Measurement of stresses in three orthogonal directions at the residual limb-prosthetic socket interface. *IEEE Transactions on Rehabilitation Engineering, 1,* 79–85.

15. Convery, P., & Buis, A. W. P. (1998). Conventional patellar-tendon-bearing (PTB) socket/stump interface dynamic pressure distributions recorded during the prosthetic stance phase of gait of a transtibial amputee. *Prosthetics and Orthotics International, 22*(3), 193–198.

16. Kane, B. J., Cutkosky, M. R., & Kovacs, G. T. (2000). A traction stress sensor array for use in high-resolution robotic tactile imaging. *Journal of Microelectromechanical Systems, 9*(4), 425–434.

17. Dabling, J. G., Filatov, A., Wheeler, J. W. (2012). Static and cyclic performance evaluation of sensors for human interface pressure measurement. In *2012 Annual International Conference of the IEEE Engineering in Medicine and Biology Society* (pp. 162–165).

18. Ruda, E. M., Sanchez, O. F. A., Mejia, J. C. H., et al. (2013). Design process of mechatronic device for measuring the stump stresses on a lower limb amputee. In *Proceedings of the 22nd Internatoinal Congress of Mechanical Engineering (COBEM 2013)*, Ribeirão Preto, Brazil (pp. 3–7).

19. Polliack, A. A., Craig, D. D., Sieh, R. C., Landsberger, S., et al. (2002). Laboratory and clinical tests of a prototype pressure sensor for clinical assessment of prosthetic socket fit. *Prosthetics and Orthotics International, 26,* 23–34.

20. Laszczak, P., Jiang, L., Bader, D. L., et al. (2015). Development and validation of a 3D-printed interfacial stress sensor for prosthetic applications. *Medical Engineering & Physics, 37*(1), 132–137.

21. Fresvig, T., Ludvigsen, P., Steen, H., et al. (2008). Fibre optic Bragg grating sensors: an alternative method to strain gauges for measuring deformation in bone. *Medical Engineering & Physics, 30*(1), 104–108.

22. Al-Fakih, E. A., Osman, N. A. A., Adikan, F. R. M., et al. (2015). Development and validation of fiber Bragg grating sensing pad for interface pressure measurements within prosthetic sockets. *IEEE Sensors Journal, 16*(4), 965–974.

23. Al-Fakih, E. A., Osman, N. A. A., Eshraghi, A., et al. (2013). The capability of fiber Bragg grating sensors to measure amputees' trans-tibial stump/socket interface pressures. *Sensors, 13*(8), 10348–10357.

24. Steege, J. W. (1988). Finite element prediction of pressure at the below-knee socket interface. In Report of ISPO Workshop on CAD CAM in Prosthetics and Orthotics.

25. Sewell, P., Noroozi, S., Vinney, J., et al. (2000). Developments in the trans-tibial prosthetic socket fitting process: A review of past and present research. *Prosthetics and Orthotics International, 24,* 97–107.

26. Zhang, M., & Mak, A. F. (1996). A finite element analysis of the load transfer between an above-knee residual limb and its prosthetic socket-roles of interface friction and distal-end boundary conditions. *IEEE Transactions on Rehabilitation Engineering, 4,* 337–346.

27. Portnoy, S., Yarnitzky, G., Yizhar, Z., et al. (2007). Real-time patient-specific finite element analysis of internal stresses in the soft tissues of a residual limb: a new tool for prosthetic fitting. *Annals of Biomedical Engineering, 35,* 120–135.
28. Zhang, M., & Roberts, C. (2000). Comparison of computational analysis with clinical measurement of stresses on below-knee residual limb in a prosthetic socket. *Medical Engineering & Physics, 22,* 607–612.
29. Zachariah, S. G., & Sanders, J. E. (2000). Finite element estimates of interface stress in the trans-tibial prosthesis using gap elements are different from those using automated contact. *Journal of Biomechanics, 33*(7), 895–899.
30. Wu, C. L., Chang, C. H., Hsu, A. T., et al. (2003). A proposal for the pre-evaluation protocol of below-knee socket design-integration pain tolerance with finite element analysis. *Journal of the Chinese Institute of Engineers, 26*(6), 853–860.
31. Lee, W. C., Zhang, M., Jia, X., et al. (2004). Finite element modeling of the contact interface between trans-tibial residual limb and prosthetic socket. *Medical Engineering & Physics, 26,* 655–662.
32. Jia, X., Zhang, M., & Lee, W. C. (2004). Load transfer mechanics between trans-tibial prosthetic socket and residual limb—dynamic effects. *Journal of Biomechanics, 37*(9), 1371–1377.
33. Portnoy, S., Yizhar, Z., Shabshin, N., et al. (2008). Internal mechanical conditions in the soft tissues of a residual limb of a trans-tibial amputee. *Journal of Biomechanics, 41*(9), 1897–1909.
34. Lacroix, D., & Patiño, J. F. R. (2011). Finite element analysis of donning procedure of a prosthetic transfemoral socket. *Annals of Biomedical Engineering, 39*(12), 2972.
35. Ramírez, J. F., & Vélez, J. A. (2012). Incidence of the boundary condition between bone and soft tissue in a finite element model of a transfemoral amputee. *Prosthetics and Orthotics International, 36*(4), 405–414.
36. Zhang, L., Zhu, M., Shen, L., et al. (2013). Finite element analysis of the contact interface between trans-femoral stump and prosthetic socket. In *2013 35th Annual International Conference of the IEEE Engineering in Medicine and Biology Society* (pp. 1270–1273).
37. Vélez Zea, J. A., Bustamante Góez, L. M., & Villarraga Ossa, J. A. (2015). Relation between residual limb length and stress distribution over stump for transfemoral amputees. *Revista EIA, 23,* 107–115.
38. Portnoy, S., Siev-Ner, I., Shabshin, N., et al. (2009). Patient-specific analyses of deep tissue loads post transtibial amputation in residual limbs of multiple prosthetic users. *Journal of Biomechanics, 42*(16), 2686–2693.
39. Lin, C. C., Chang, C. H., Wu, C. L., et al. (2004). Effects of liner stiffness for trans-tibial prosthesis: A finite element contact model. *Medical Engineering & Physics, 26*(1), 1–9.
40. Tang, J., McGrath, M., Laszczak, P., et al. (2015). Characterisation of dynamic couplings at lower limb residuum/socket interface using 3D motion capture. *Medical Engineering & Physics, 37*(12), 1162–1168.
41. Sanders, J. E., Greve, J. M., Mitchell, S. B., et al. (1998). Material properties of commonly-used interface materials and their static coefficients of friction with skin and socks. *Journal of Rehabilitation Research and Development, 35,* 161–176.
42. Zhang, M., & Mak, A. F. T. (1999). In vivo friction properties of human skin. *Prosthetics and Orthotics International, 23*(2), 135–141.
43. Derler, S., & Gerhardt, L. C. (2012). Tribology of skin: review and analysis of experimental results for the friction coefficient of human skin. *Tribology Letters, 45*(1), 1–27.
44. Zhang, M., Lord, M., Turner-Smith, A. R., et al. (1995). Development of a non-linear finite element modelling of the below-knee prosthetic socket interface. *Medical Engineering & Physics, 17,* 559–566.
45. Valanis, K. C., & Landel, R. F. (1967). The strain-energy function of a hyperelastic material in terms of the extension ratios. *Journal of Applied Physics, 38*(7), 2997–3002.
46. Attard, M. M., & Hunt, G. W. (2004). Hyperelastic constitutive modeling under finite strain. *International Journal of Solids and Structures, 41,* 5327–5350.
47. Natali, A. N., Carniel, E. L., Pavan, P. G., et al. (2006). Hyperelastic models for the analysis of soft tissue mechanics: definition of constitutive parameters. In *The First IEEE/RAS-EMBS International Conference on Biomedical Robotics and Biomechatronics* (pp. 188–191).

48. Marckmann, G., & Verron, E. (2006). Comparison of hyperelastic models for rubber-like materials. *Rubber Chemistry and Technology, 79,* 835–858.
49. Portnoy, S., Siev-Ner, I., Shabshin, N., et al. (2011). Effects of sitting postures on risks for deep tissue injury in the residuum of a transtibial prosthetic-user: A biomechanical case study. *Computer Methods in Biomechanics and Biomedical Engineering, 14,* 1009–1019.
50. Portnoy, S., Siev-Ner, I., Yizhar, Z., et al. (2009). Surgical and morphological factors that affect internal mechanical loads in soft tissues of the transtibial residuum. *Annals of Biomedical Engineering, 37*(12), 2583.
51. Silver-Thorn, M. B. (1999). In vivo indentation of lower extremity limb soft tissues. *IEEE Transactions on Rehabilitation Engineering, 7,* 268–277.
52. Tönük, E., & Silver-Thorn, M. B. (2004). Nonlinear viscoelastic material property estimation of lower extremity residual limb tissues. *IEEE Transactions on Neural Systems and Rehabilitation Engineering, 126,* 289–300.

3D Brain Deformation in Cadaveric Specimens Compared to Healthy Volunteers Under Non-injurious Loading Conditions

Andrew K. Knutsen, Philip V. Bayly, John A. Butman, and Dzung L. Pham

Abstract Measurements of how the brain deforms in response to head impact are often obtained in human cadavers because loading conditions can be investigated that are on the order of those that could lead to injury. How the deformation response of the postmortem human brain differs from the live human brain remains unanswered. In this study, we used tagged magnetic resonance imaging (MRI) to measure brain deformation in two human cadavers in response to non-injurious head impact. We then compared these strain fields to those obtained in 10 healthy volunteers (HV) during the same neck rotation under similar loading conditions. Our results showed a number of similarities, such as a similar magnitude of the largest maximum principal strain values and high values within the cortex, and some differences, such as high strain concentrations near the falx, lower median strain values at each time point, and an increased frequency of oscillation in cadavers. These initial results provide insight into potential differences in the deformation response of the brain in cadavers compared to HVs, though additional experiments are needed to further investigate the effect of age and atrophy on the deformation response.

Keywords Tagged MRI · Brain deformation · Cadaver · In vivo · Strain

A. K. Knutsen (✉) · D. L. Pham
Center for Neuroscience and Regenerative Medicine, The Henry M Jackson Foundation,
Bethesda, MD, USA
e-mail: andrew.knutsen@nih.gov

P. V. Bayly
Department of Mechanical Engineering and Materials Science, Washington University in St.
Louis, St. Louis, MO, USA

J. A. Butman
Clinical Center, National Institutes of Health, Bethesda, MD, USA

© The Author(s), under exclusive license to Springer Nature Switzerland AG 2021   113
K. Miller et al. (eds.), *Computational Biomechanics for Medicine*,
https://doi.org/10.1007/978-3-030-70123-9_9

1 Introduction

Traumatic brain injury (TBI) is a common health condition, with over 2.5 million emergency department visits per year in the United States, that can lead to chronic symptoms, disability, and death [1]. Penetrating forms of TBI can lead to more severe injuries, though skull fracture is not necessary for TBI. Concussion and other non-penetrating forms of TBI, such as diffuse axonal injury, can result from the rapid deformation of brain tissue due to head impact or acceleration [2]. An understanding of the spatiotemporal pattern of brain deformation in response to known head accelerations is necessary to predict mechanisms of injury and to design protective strategies to prevent injury.

Measurements of brain deformation during head impact obtained in human cadavers provide useful benchmark data for comparison to computational models of TBI. Multiple studies have provided such data. Hardy et al. used bi-planar x-ray to measure brain displacement during head impact in cadavers [3, 4]. Follow-on studies provided coarse estimates of planar strain from clusters of the implanted neutral density trackers [5, 6]. Some more recent experiments used sonomicrometry to measure brain deformation in cadavers during rotational head impacts of varying severity [7, 8].

The primary advantage of these cadaveric experiments is that they can obtain accurate measures of brain displacement during impact severities relevant to head injury. However, the use of cadaver data to drive the development of protective equipment and computational simulations of TBI also has some disadvantages. One challenge is that the displacements have been measured in a limited number of locations, providing sparse and inaccurate estimates of local strain. Additionally, the postmortem brain is stiffer than the live brain [9, 10]. How these changes affect brain deformation during injury remains poorly understood.

Tagged magnetic resonance imaging (MRI) has been used to obtain dense measurements of 3D brain deformation during very mild, non-injurious head impacts in healthy volunteers (HV) [11, 12] and has also been applied to measure 3D deformation in four axial slices in a human cadaver [13]. Because tagged MRI creates contrast that can be used to non-invasively measure the deformation of soft tissue without using ionizing radiation, it is an ideal methodology for characterizing the biomechanical response of the brain in both live and postmortem human specimens under similar loading conditions. In this study, we used tagged MRI to measure 3D brain deformation in response to a mild deceleration after neck rotation in two human cadavers. Measurements were obtained over most of the brain at high spatial resolution, allowing dense measurement of strain. We then compared these measurements with those obtained in 10 HVs during the same head motion.

2 Methods

Head and neck specimens from two human cadavers transected at approximately the third to sixth cervical vertebra, were obtained from the Maryland State Anatomy Board and imaged within 72 h of death. The first was female, age 91 (C1), and the second was male, age 79 (C2). The motion of interest was a neck rotation about the inferior-superior axis of the head towards the left side of the specimen. A custom device was used to generate reproducible head motion (Fig. 1a). This device has been described previously in studies of healthy volunteers [11, 12, 14–16] and of a cadaver [13]. An offset counterweight accelerated the specimen through 32 degrees of rotation, at which point the counterweight impacted a padded stop, which caused rapid deceleration of ~250 rad/s^2. Angular position (θ), velocity (ω), and acceleration (α) as a function of time were measured and recorded using an MRI-compatible angular position encoder (MICRONOR, Camarillo, CA, USA).

We used a Siemens 3 T Biograph mMR (Erlangen, Germany) scanner with a 16-channel head receive coil to acquire a structural T1-weighted MPRAGE with the following image parameters: echo time (TE) = 3.03 ms, repetition time (TR) = 2530 ms, inversion time (TI) = 1100 ms, field of view (FOV) = 256 × 256 × 176 mm,

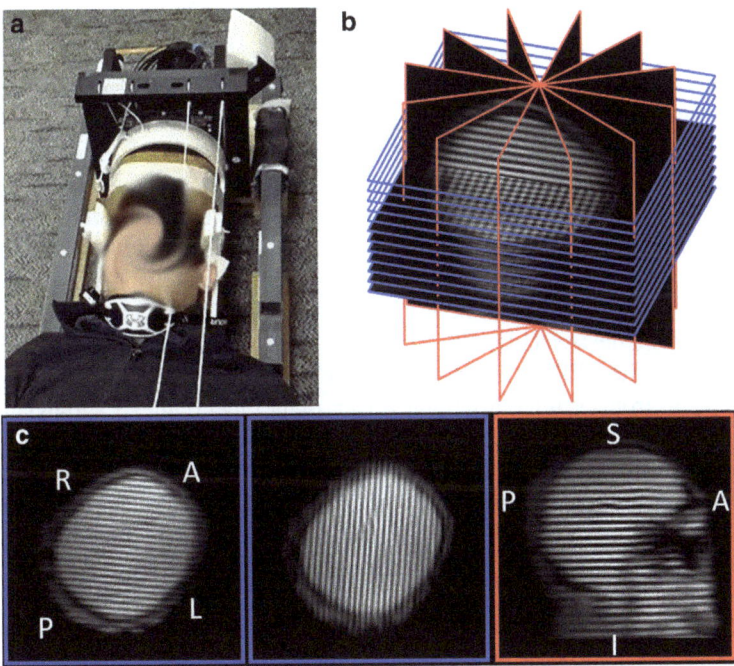

Fig. 1 a MRI-compatible head rotation device used to induce repeatable mild head impact during neck rotation. **b** Multi-slice tagged MRI acquisition strategy to cover the entire brain. **c** Tagged MRI data showing the three tag orientations

matrix = 256 × 256 × 176. Tagged MRI was used to obtain non-invasive images of tissue deformation over time. The acquisition protocol was similar to what has been described previously for healthy volunteers [11, 12] with the following imaging parameters: TR/TE = 3.01/1.67 ms, tag spacing = 6 mm, segments = 6, FOV = 240 × 240 mm, matrix = 24 × 160 (reconstructed to 160 × 60), slice thickness = 8 mm with a 2 mm gap between slices, temporal resolution = 18 ms, number of image frames = 10.

To observe motion in three dimensions, a series of tagged MR images were acquired in the axial plane, with tag lines along the x- and y-axes, and in planes perpendicular to axial (rotated about the center of the brain like spokes on a wheel) with tag lines along the z-axis (Fig. 1b). Multiple slices were acquired to cover the majority of the brain: 13 axial slices for C1, 11 axial slices for C2, and six orthogonal slices for both cadavers. Sample tagged images are shown in Fig. 1c. Based on the spatial and temporal resolution of the acquisition, the motion was repeated four times per image slice. As such, to acquire all the image data, the motion was repeated 128 times for C1 and 112 times for C2.

Estimates of displacement and Lagrangian strain were computed from the tagged MRI data using the harmonic phase finite element (HARP-FE) method [11]. The processing steps match those used in a previous study of healthy volunteers [12]. Briefly, tagged MR images for frames 2–10 were rigidly registered to frame 1 using the Advanced Normalization Toolkit v2.1 [17]. The registered images were then spatially interpolated to an isotropic grid at 1.5 mm resolution in MATLAB R2016a (Mathworks, MA, USA) using cubic spline interpolation. For each tag direction, following a 3D Fourier transform for each frame, a bandpass filter with a radius of nine voxels and centered at the spatial frequency of the tag lines was applied; the filter size was selected from previous studies [12, 15, 16]. Harmonic phase (HARP) images were created by computing the phase of the inverse Fourier transform of the filtered spatial frequency images. The HARP signal remains constant as the tissue deforms and can be used to track tissue motion (e.g., [18]). Here, we used the difference in the HARP images between each frame and frame 1 to create a body force that drives a 3D finite element model (see [11]). The algorithm, termed HARP-FE, was implemented as a plugin to the FEBio software [19]. The output from HARP-FE consists of displacements from the reference frame to each of the other frames. Lagrangian strain tensors were computed in MATLAB using a local fitting approach [12]. Eigenvalues and eigenvectors were computed from the Lagrangian strain tensors. The maximum principal strain (MPS) was assigned to be the first eigenvalue. For each frame, the 50th percentile (MPS50) and 95th percentile (MPS95) of the distribution within the brain tissue were assessed for each data set.

For comparison, data from 10 HVs (five male, five female, ages 24–40) were used [12]. The HVs provided informed consent under a protocol approved by the National Institutes of Health CNS Institutional Review Board, and all experiments were conducted in the Clinical Center at the National Institutes of Health using the protocol described above, except that a tag spacing of 8 mm was used.

The MPS values in vivo show a strong linear relationship with angular velocity at impact (ω_{max}) [12]. To account for differences in the loading conditions and compare

MPS across subjects, MPS50 and MPS95 values for all data sets were normalized using a scale factor, which was given by the ratio of the average of ω_{max} of all HVs to ω_{max} for each subject.

3 Results

Figure 2 shows the traces of ω (Fig. 2a) and α (Fig. 2b) versus time for 128 repetitions in specimen C1. This illustrates the repeatability of the measurements obtained. The average of the maximum of ω (ω_{max}) versus α (α_{max}) is shown in Fig. 2c, which provides a range of the loading conditions observed for both the cadaver and HV data sets. The average of ω_{max} in the HV data was used to normalize the MPS values.

The 50th and 95th percentile of MPS was computed as a function of time for both cadaver data sets are shown in Fig. 3. For comparison, the mean ± standard error or the HV data sets are also provided. The largest values for MPS50 and MPS95 are seen in the third frame (~27 ms after ω_{max}) for both the cadavers and HVs. Both data sets show damped oscillations, though the frequency of oscillation appears higher for the two cadaver data sets. After normalization based on ω_{max}, the MPS50 values are slightly lower than those in the HV data sets, while the MPS95 are of similar magnitude.

Figure 4 contains orthogonal views of MPS versus time for each of the cadaver data sets, and shows a complex spatial pattern of MPS over time. At the time of peak deformation (frame 3), large values of MPS were seen in the frontal cortex (seen in all orientations) and near the falx (note the coronal orientation). Additionally, MPS

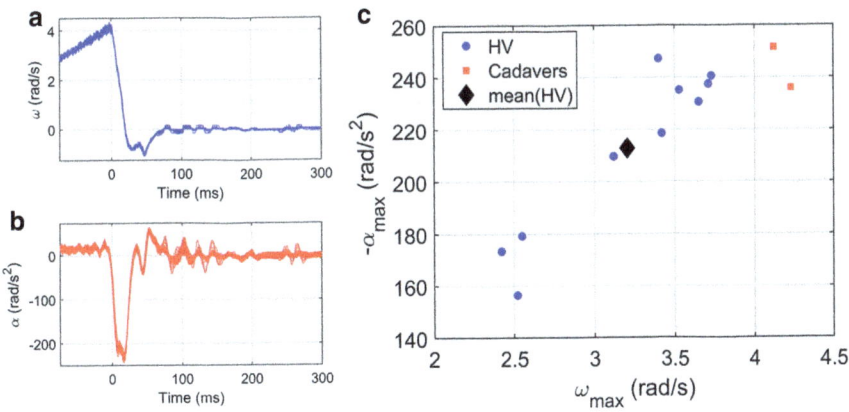

Fig. 2 This figure shows the recorded traces of **a** angular velocity (ω) and **b** angular acceleration (α) versus time for 128 repetitions obtained in Cadaver 1. The time zero corresponds to the maximum ω (just prior to impact). **c** shows α vs ω for the HV and cadaver data sets. The average value of the HV data was used to normalize the metrics of maximum principal strain

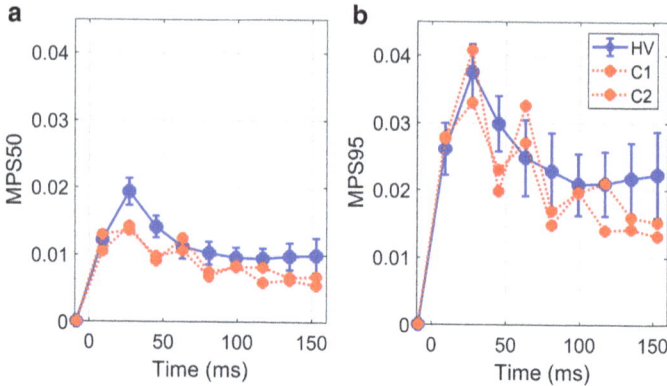

Fig. 3 This figure shows normalized values of the **a** MPS50 (50th percentile of MPS) and **b** MPS95 (95th percentile of MPS) versus time for each of the cadaver data sets and the HVs (mean ± SEM). Zero on the time axis was set to the time when ω_{max} occurred

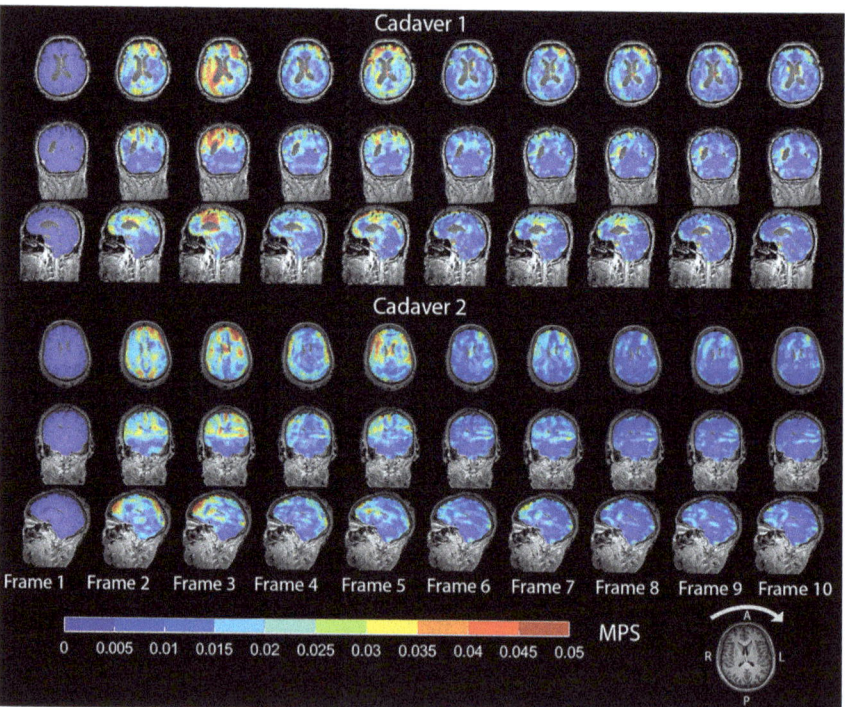

Fig. 4 Axial, coronal, and sagittal views of the MPS versus time for both cadaver data sets. Frame 1 was acquired ~9 ms prior to impact, and the temporal resolution is 18 ms per frame. The arrow over the anatomical image in the bottom right corner indicates the direction of neck rotation; A = anterior, P = posterior, R = right, L = left

values appeared noticeably larger in the right hemisphere in Cadaver 1 (axial and coronal orientations), which was opposite the location of impact.

Additional information can be obtained from the orientation of the deformation field at the time of peak deformation (Fig. 5). To illustrate this, spheres in every other voxel in the reference configuration were mapped to ellipsoids in the deformed configuration and overlaid on the subject anatomy for the two cadaver data sets and two of the 10 HV data sets. The color scale shows the value of MPS. We note that the MPS values shown in Fig. 5 are not scaled by ω_{max}. The amount of deformation visualized by the ellipsoids was scaled by a factor of 10 to exaggerate the deformation that occurred.

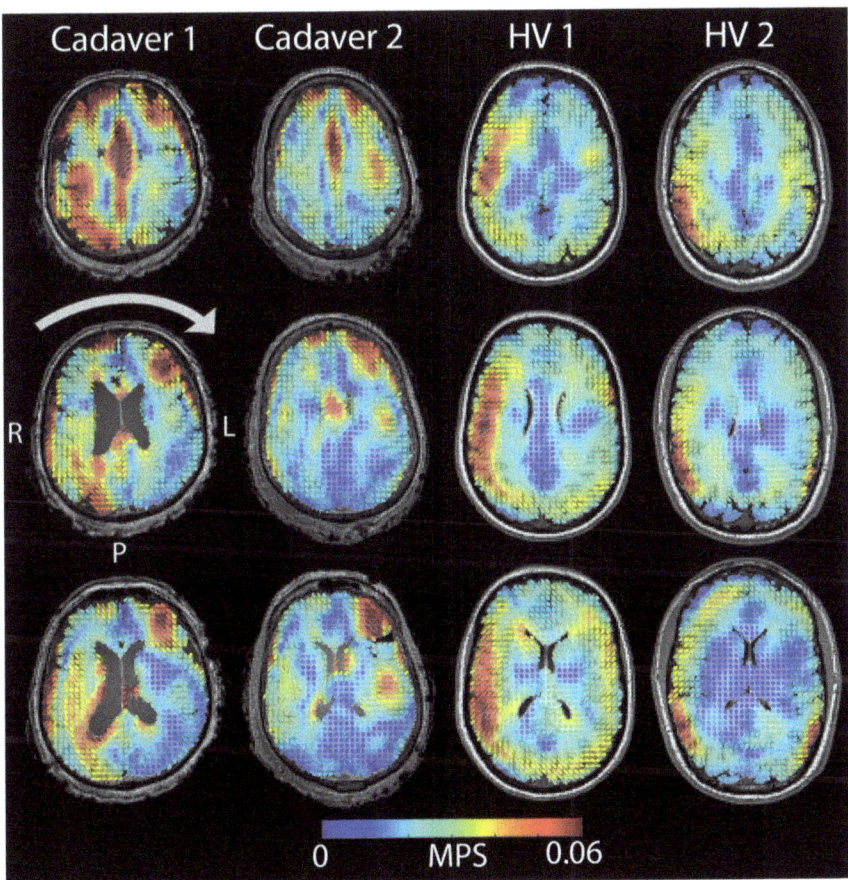

Fig. 5 Axial view of MPS at the time of peak deformation for the two cadaver data sets and two of the 10 HV data sets. To visualize the magnitude and orientation of the deformation, spheres were mapped from the reference to deformed configuration. The block arrow indicates the direction of rotation. P = posterior, R = right, L = left

As noted above, large values of MPS were seen in the peripheral cortex and near the falx in both cadaver data sets. While larger values of MPS were also seen in the cortex for the HV data sets, the effect of the falx was less obvious on the deformation response. The ellipsoids at the outer bounds of the cortex were often oriented along the direction of rotation in both the cadaver and HV data sets. For the HV data sets, lower MPS values tended to occur near the center of rotation, while for the cadavers, lower MPS values were observed in posterior regions, especially on the left hemisphere.

4 Discussion

The measured MPS values in cadavers showed multiple similarities and some differences compared to data acquired from the HVs. After normalization, the median magnitude of deformation (MPS50) was slightly lower in the two cadavers than the HV data sets. This could be due to increased stiffness of the brain tissue that occurs postmortem [9, 10]. The magnitude of the largest strain values (MPS95) was similar between the cadavers and HVs. High MPS values were observed in the peripheral cortex near the convexities for all data sets, while the cadavers showed concentrations of high MPS values near the falx. Additionally, both the MPS50 and MPS95 values oscillated at a faster rate in the cadavers than in the HVs.

Several factors could have affected these observations. First, we have only acquired data in two cadavers to date. These cadavers were considerably older than the HVs that we have imaged (79 and 91 vs. 24–40 years of age). Aging leads to increased ventricles and atrophy of the cortex (e.g., [20]). Atrophy of the cortex and increased ventricle size could reasonably lead to more space for the brain to deform unimpeded. As an example, C1 had larger ventricles than C2 and also had larger strains at the time of peak deformation (Figs. 4 and 5). Additionally, the increase in the stiffness of the postmortem human subject brain tissue could explain the higher rate of oscillation during damping. It will be important to identify the effect due to age versus differences between living and postmortem brain tissue. In future studies, we will compare cadaver results to strains from healthy volunteers with brains that more closely match the cadavers in terms of age and structure, including the degree of atrophy. We also plan to use computational modeling to look at the effect of atrophy and ventricle size on the brain's deformation response.

For this study, we used the MPS because it is a scalar quantity that represents the magnitude of deformation and because it is commonly used in studies of brain deformation (e.g., [21, 22]). Additional quantities, such as maximum shear strain or the full tensor information could have been used as well. MPS50 provides an estimate of the median amount of deformation, and MPS95 provides an estimate of the peak amount of deformation while avoiding potential numeric artifacts using the maximum value [23].

The loading conditions used in this study were well below those that could lead to injury in the live human brain. A study by Funk et al. measured head accelerations of up to 31 g and 2888 rad/s^2 in HVs under a range of activities, such as a soccer ball impact to the forehead, voluntary head shaking, plopping down on a chair, and jumping off of a step [24]. For comparison, the peak angular accelerations measured in this study were approximately an order of magnitude lower. While this limitation holds for HVs, it does not for cadavers. In future experiments, we plan to test the cadavers under a larger range of loading conditions and using magnetic resonance elastography to measure viscoelastic material properties [25]. The maximum temporal resolution that can be obtained on our scanner is ~2–3 ms per frame, which should be sufficient to capture the deformation response under increased loading.

5 Conclusions

In this study, we used tagged MRI to obtain 3D measurements of Lagrangian strain in two human cadaver head and neck specimens. The use of tagged MRI allowed for dense, high resolution estimates of strain fields throughout the brain, which showed complex spatial patterns of MPS that varied over time. These strain fields were compared to those obtained in 10 HVs. These results showed similarities in the magnitude of the peak strains (MPS95) and that high strains were observed in the cortex in all data sets; differences included decreased strain response (MPS50), high strain concentrations, and an increased frequency of oscillation in the cadavers compared to HVs. Future experiments will attempt to clarify the effect of age, atrophy, and the use of postmortem brain tissue on the deformation response of the brain.

Acknowledgements We acknowledge funding from the National Institute of Neurological Disorders and Stroke (R01/R56 NS055951 and U01 NS112120), the Department of Defense in the Center for Neuroscience and Regenerative Medicine, and the Intramural Research Program of the National Institutes of the Health.

References

1. Taylor, C. A., Bell, J. M., Breiding, M. J., et al. (2017). Traumatic brain injury-related emergency department visits, hospitalizations, and deaths-United States, 2007 and 2013. *MMWR Surveillance Summaries, 66*(9), 1–16.
2. Meaney, D. F., Smith, D. H., Shreiber, D. I., et al. (1995). Biomechanical analysis of experimental diffuse axonal injury. *Journal of Neurotrauma, 12*(4), 689–694.
3. Hardy, W. N., Foster, C. D., Mason, M. J., et al. (2001). Investigation of head injury mechanisms using neutral density technology and high-speed biplanar X-ray. *Stapp Car Crash Journal, 45*, 337–368.
4. Hardy, W. N., Mason, M. J., Foster, C. D., et al. (2007). A study of the response of the human cadaver head to impact. *Stapp Car Crash Journal, 51*, 17–80.
5. Zhou, Z., Li, X., Kleiven, S., et al. (2019). Brain strain from motion of sparse markers. *Stapp Car Crash Journal, 63*, 1–27.
6. Zhou, Z., Li, X., Kleiven, S., et al. (2018). A reanalysis of experimental brain strain data: implication for finite element head model validation. *Stapp Car Crash Journal, 62*, 293–318.

7. Alshareef, A., Giudice, J. S., Forman, J., et al. (2018). A novel method for quantifying human in situ whole brain deformation under rotational loading using sonomicrometry. *Journal of Neurotrauma, 35*(5), 780–789.

8. Alshareef, A., Giudice, J. S., Forman, J., et al. (2020). Biomechanics of the human brain during dynamic rotation of the head. *Journal of Neurotrauma.*

9. Weickenmeier, J., Kurt, M., Ozkaya, E., et al. (2018). Brain stiffens post mortem. *Journal of the Mechanical Behavior of Biomedical Materials, 84,* 88–98.

10. Guertler, C. A., Okamoto, R. J., Schmidt, J. L., et al. (2018). Mechanical properties of porcine brain tissue in vivo and ex vivo estimated by MR elastography. *Journal of Biomechanics, 69,* 10–18.

11. Gomez, A. D., Knutsen, A. K., Xing, F., et al. (2019). 3-D measurements of acceleration-induced brain deformation via harmonic phase analysis and finite-element models. *IEEE Transactions on Biomedical Engineering, 66*(5), 1456–1467.

12. Knutsen, A. K., Gomez, A. D., Gangolli, M., et al. (In Press). In vivo estimates of axonal stretch and 3D brain deformation during mild head impact. *Brain Multiphysics.*

13. Knutsen, A. K., Wang, W. T., McEntee, J. E., et al. (2012) Using tagged MRI to quantify the 3D deformation of a cadaver brain in response to angular acceleration. In S. N. York (Ed.), *Computational Biomechanics for Medicine: Models, Algorithms and Implementation* (pp. 169–183).

14. Sabet, A. A., Christoforou, E., Zatlin, B., et al. (2008). Deformation of the human brain induced by mild angular head acceleration. *Journal of Biomechanics, 41*(2), 307–315.

15. Chan, D. D., Knutsen, A. K., Lu, Y. C., et al. (2018). Statistical characterization of human brain deformation during mild angular acceleration measured in vivo by tagged magnetic resonance imaging. *Journal of Biomechanical Engineering, 140*(10).

16. Knutsen, A. K., Magrath, E., McEntee, J. E., et al. (2014). Improved measurement of brain deformation during mild head acceleration using a novel tagged MRI sequence. *Journal of Biomechanics, 47*(14), 3475–3481.

17. Avants, B. B., Tustison, N. J., Song, G., et al. (2011). A reproducible evaluation of ANTs similarity metric performance in brain image registration. *Neuroimage, 54*(3), 2033–2044.

18. Liu, X., & Prince, J. L. (2010). Shortest path refinement for motion estimation from tagged MR images. *IEEE Transactions on Medical Imaging, 29*(8), 1560–1572.

19. Maas, S. A., Ellis, B. J., Ateshian, G. A., et al. (2012). FEBio: finite elements for biomechanics. *Journal of Biomechanical Engineering, 134*(1), 011005.

20. Fjell, A. M., Walhovd, K. B., Fennema-Notestine, C., et al. (2009). One-year brain atrophy evident in healthy aging. *Journal of Neuroscience, 29*(48), 15223–15231.

21. Anderson, E. D., Giudice, J. S., Wu, T., et al. (2020). Predicting concussion outcome by integrating finite element modeling and network analysis. *Frontiers in Bioengineering and Biotechnology, 8,* 309.

22. Wu, S., Zhao, W., Ghazi, K., et al. (2019). Convolutional neural network for efficient estimation of regional brain strains. *Scientific Reports, 9*(1), 17326.

23. Panzer, M. B., Myers, B. S., Capehart, B. P., et al. (2012). Development of a finite element model for blast brain injury and the effects of CSF cavitation. *Annals of Biomedical Engineering, 40*(7), 1530–1544.

24. Funk, J. R., Cormier, J. M., Bain, C. E., et al. (2011). Head and neck loading in everyday and vigorous activities. *Annals of Biomedical Engineering, 39*(2), 766–776.

25. Hiscox, L. V., Johnson, C. L., Barnhill, E., et al. (2016). Magnetic resonance elastography (MRE) of the human brain: technique, findings and clinical applications. *Physics in Medicine & Biology, 61*(24), R401–R437.

Index

© The Editor(s) (if applicable) and The Author(s), under exclusive license
to Springer Nature Switzerland AG 2021
K. Miller et al. (eds.), *Computational Biomechanics for Medicine*,
https://doi.org/10.1007/978-3-030-70123-9

Lightning Source UK Ltd.
Milton Keynes UK
UKHW020611250722
406326UK00002B/25